Sustainable Finance

Sustainable Finance is a concise and authoritative reference series linking research and practice. It provides reliable concepts and research findings in the ever growing field of sustainable investing and finance, SDG economics and Leadership with the declared commitment to present the theories, methods, tools and investment approaches that can fulfil the United Nations Sustainable Development Goals and the Paris Agreement COP 21/22 alongside with de-risking assets and creating triple purpose solutions that ensure the parity of profit, people and planet through choice architecture passion and performance. The series addresses market failure, systemic risk and reinvents portfolio theory, portfolio engineering as well as behavioural finance, financial mediation, product innovation, shared values, community building, business strategy and innovation, exponential tech and creation of social capital. Sustainable Finance and SDG Economics series helps to understand keynotes on international guidelines, guiding accounting and accountability principles, prototyping new developments in triple bottom line investing, cost benefit analysis, integrated financial first plus impact first concepts and impact measurement. Going beyond adjacent fields (like accounting, marketing, strategy, risk management) it integrates the concept of psychology, innovation, exponential tech, choice architecture, alternative economics, blue economy shared values, professions of the future, leadership, human and community development, team culture, impact, quantitative and qualitative measurement, Harvard Negotiation, mediation and complementary currency design using exponential tech and ledger technology. Books in the series contain latest findings from research, concepts for implementation, as well as best practices and case studies for the finance industry.

Stefan Brunnhuber

The Third Culture

The Impact of AI on Knowledge, Society and Consciousness in the 21st Century

Springer

Stefan Brunnhuber
Trustee of the World Academy of Art and Science
Mittweida, Germany

ISSN 2522-8285 ISSN 2522-8293 (electronic)
Sustainable Finance
ISBN 978-3-031-48112-3 ISBN 978-3-031-48113-0 (eBook)
https://doi.org/10.1007/978-3-031-48113-0

This Springer imprint is published by the registered company Springer Nature Switzerland AG
The registered company address is: Gewerbestrasse 11, 6330 Cham, Switzerland

Paper in this product is recyclable.

I dedicate this book to
my wife
Stephanie.
I am grateful for her endless
compassion, love
commitment and spirit.

Preface

The traditional 'two cultures' view (S. P. Snow) distinguishes between the sciences and the humanities. The *sciences* include physics, chemistry, biology, mathematics and engineering. Their main interest is in exploring natural laws and applying them to real-life problems. Meanwhile, the main goal of the *humanities*, including disciplines such as philosophy, history, linguistics and qualitative sociology and psychology, is to interpret the world and attain a deeper understanding of our history, cultural activities and psyche. The wisdoms that these two cultures offer—respectively, *explanation* and *understanding*—are separated from each other, with little to no interaction or mutual comprehension. However, over the past two decades, and for the first time in human history, a new, third culture has appeared on the historical battlefield. This new culture, rooted in new technologies, not only pursues its own form of rationality but also supports advances in the original two cultures, which will further loop back into society, doubling the world in digital form and eventually deepening and expanding our individual and collective consciousness so that we can see more and do better. Furthermore, research and development are destined to become truly transdisciplinary, paving the way for a form of integrated knowledge that we could call 'one science'. These new technologies will reveal the interconnectedness, vulnerability, interdependency and boundaries of the world and fundamentally redefine the human species' position in the twenty-first century: not a conductor leading the orchestra, but a single string player within it. We are entering a second Renaissance, in which these new technologies become powerful integrators. A second Renaissance that will redefine transhumanism, the concept of singularity, the garbage in, garbage out effect, the black box dilemma and much more besides. And that will eventually give rise to new forms of consciousness based not on biochemical signals, but on copper wires and lithium chips. These new machine intelligences will change the world and force us to realise: we are no longer alone.

Mittweida, Germany
December 2023

Stefan Brunnhuber

Contents

1 Finding the Narrative: Shifting East 1
1.1 Introduction 1
1.2 The Questions behind It All 6
1.3 Respecting Boundaries 8

2 Finding Potential Integrators 13
2.1 A Reformed Financial Sector 14
2.2 Altered States of Mind 24
2.3 'Metastability' and the Logic of Fractals 35

3 The 'Two Cultures' Debate and the Logic of Scientific Revolutions 39
3.1 The Two Cultures 39
3.2 The Structure of Any Scientific Revolution 40
3.3 Understanding Complexity 2.0 41

4 Towards Three Cultures 45
4.1 Examples and Best Practices 46
4.2 The Ghost in the Machine 52

5 On Consciousness: The Evolving Mind 55
5.1 Defining Consciousness 56
5.2 On Science and *Sapientia* 60
5.3 The Inverse Pyramid 62

6 Towards a Third Culture 67
6.1 The Phenomenology of a New Technology 67
6.2 Garbage In, Garbage Out and the Black Box Effect 69
6.3 Filling the Gap: The Technological Singularity 71
6.4 The Broader Spectrum of Our Consciousness 76

7 Being Human in the Twenty-First Century 81
7.1 The Deficient Species and Its Crutch 81
7.2 The Middle Dimension and the Gap 88
7.3 The Ladder and the String Player 91

8 Questionary: An Adjusted Turing Test 95

9 The Dawn of a New Integral Wisdom 97

Bibliography 101

About the Author

Stefan Brunnhuber is Medical Director, Chief Medical Officer and Professor for Sustainability and Finance at Mittweida University of Applied Sciences (Germany). He is a member of the European Academy of Sciences and Arts (EASA), a trustee of the World Academy of Art and Science (WAAS) and a full member of the Club of Rome. He was scientific advisor to the EU Commission (2009), member of the Lancet Commission (2021–2022), founding member of Alma Mater Europeae (2011) and senator (elected 2015–2020) of the European Academy of Sciences and Arts (EASA). He has served since 2022 on the German Federal Government's Sustainable Finance Advisory Committee.

List of Figures

Fig. 1.1	Different stages of societal aggregation: horizontal and vertical evolution	9
Fig. 1.2	The dawn of a new era: towards integral consciousness	12
Fig. 2.1	The money system serves as an attractor and integrator	14
Fig. 2.2	The traditional way to do it—end-of-pipe, redistributive approach	16
Fig. 2.3	Overcoming polarities: triangulating the system	20
Fig. 2.4	Rethinking finance: CBDCs and CBCSs can provide the liquidity and leverage to fund, manage and hedge our commons	24
Fig. 2.5	Thinking outside the square box—the nine dots puzzle	25
Fig. 2.6	Push and pull factors leading to altered, more integral states of mind	27
Fig. 2.7	Non-curricular building blocks for an altered mindset	30
Fig. 2.8	The circle of creativity	32
Fig. 2.9	The human cognitive reserve—the larger the better	33
Fig. 2.10	Towards metastability: altered states of consciousness that integrate opposites	36
Fig. 2.11	The logic of fractals—a scale-independent isomorphic state	36
Fig. 3.1	From risks to uncertainties to unknown unknowns	43
Fig. 4.1	Not breaking but expanding the hermeneutic circle	49
Fig. 5.1	The seven-stage inverse pyramid: from observation to wisdom	63
Fig. 5.2	An updated version of Descartes's vision: symbolic-representational—constructivist—connectivist	64
Fig. 6.1	An uncertain future: will AI and deep learning follow an exponential or a logistic curve?	69
Fig. 6.2	The unknown collective mirror effect	70
Fig. 6.3	Matching the human and the digital mind: the black box and garbage in, garbage out problems	71

Fig. 6.4 (**a**) Breakdown of nine forms of human intelligence compared with the average. (**b**) Breakdown of different forms of human IQ complemented by their digital counterparts (level 2, see explanation in the text). (**c**) Digital IQ replaces and extends human capacities. The arrows represent digital augmentation on the one side and replacement on the other 73
Fig. 6.5 The three levels of technological singularity 75
Fig. 6.6 The developmental logic of AI and datafication 76
Fig. 6.7 One Science 2.0: the ‘third culture’ and the new scientific revolution 77
Fig. 6.8 Metastability and triangulation 78
Fig. 7.1 (**a**) Animals are fully adapted to their environment. (**b**) Humans are a deficient species, never fully adapted to nature (blue) and always requiring crutches (yellow) to fill the gap 87
Fig. 7.2 Timeline: as the gap gets bigger, the more that is needed to bridge it 90
Fig. 7.3 Humans at the centre of the universe (**a**); evolution not as a ladder (**b**) but as overlapping, asymmetric circles (**c**), with humans playing the marginal, fragile but essential role of a string player 92

List of Tables

Table 2.1	Money is not neutral—six elements of the current system that are damaging our future	15
Table 2.2	Outline of a green transition plan	22
Table 2.3	Differences between the exoteric and esoteric aspects of religion	27
Table 3.1	One culture and the two cultures (science and humanities): explanation and understanding	40
Table 4.1	Digitalisation as the third culture: transcending, augmenting and integrating 'understanding' and 'explanation'	53
Table 5.1	The 8 Ss for the formation of any consciousness	59
Table 6.1	The triple strategy for survival: storage, replication and self-improvement	68
Table 7.1	A combination of more (A) and less (B) specific features determines what it is to be human	85
Table 8.1	Are you speaking to a human? And is your interlocutor conscious or not?	96

List of Boxes

Box 1.1 The emergence of new technologies 10
Box 2.1 The web without a weaver paradox: DeFi and CeFi 19
Box 2.2 A case study: rescuing the Amazon with a CBCS 21
Box 2.3 What we should avoid: self-exemption 34
Box 2.4 Mind-altering psychedelic drugs 34
Box 3.1 The emergence of a third culture: a first look 41
Box 4.1 Dealing with the shadows 51
Box 4.2 Two sides of the coin 54
Box 5.1 The universality of hardware-independent consciousness 56
Box 5.2 The many possible foundations of consciousness 60
Box 5.3 Science and *sapientia* 63
Box 6.1 The features of intelligence 72
Box 6.2 The third culture 78
Box 7.1 Complexity explained 86
Box 7.2 On anthropomorphism 89
Box 7.3 The ladder and the string player 93
Box 8.1 A revised Turing test 96
Box 9.1 The five Ps 100

Chapter 1
Finding the Narrative: Shifting East

1.1 Introduction

This is not yet another book about artificial intelligence (AI). It is a book about the impact of a new technology on our minds, our consciousness, our society and our common future. New technologies equal new perceptions, new practices and new understandings. As we invent new tools, we recreate and mirror ourselves in their image. This was true of the discovery of fire and the invention of the wheel. It was true of Newtonian mechanics, which views the world as a machine, and systems theory, which views everything as a network. And it will be true of AI, deep learning and datafication too. The underlying question I seek to address is: *what is it to be human in the twenty-first century, at the dawn of new technologies—AI, big data, the Internet of Things?* What makes the human species distinctive and successful, and what gives us a selection advantage, has less to do with individual competitiveness, sophisticated tool use, walking upright and abstract, analytical forms of thinking than it does with our capacity to tell each other credible stories. In most cases, these stories do not refer to the objective natural world around us, but rather to a second, human-made, cultural reality. These narratives are mainly about God, death, technology, laws of nature, money, power and politics.[1] It is precisely the shared belief in these fictional stories that enables humans to coordinate and collaborate on a large scale. Human history has shown that it is better to have a false story than no story at all. Narratives, even when they are false, serve to stabilise both the individual and the collective psyche. They operate like a crutch, supporting a human species that is never fully adapted to nature. Only time will tell whether the fictional story about the future validates current human activities, and whether it is self-fulfilling or self-

[1] The more expansive a narrative becomes over time, the more powerful it is. This apparent paradox derives from the fact that until an alternative story is found, we are forced to give the victims and losses resulting from this initial narrative a sense of ultimate purpose. This is also true of powerful narratives concerning the impact of new technologies on human societies.

S. Brunnhuber, *The Third Culture*, Sustainable Finance,
https://doi.org/10.1007/978-3-031-48113-0_1

negating. Moreover, as long as we do not have an alternative story to tell ourselves, or a different mental frame to explain the world around us, any threat to our sense of coherence will override scientific knowledge. In short, our current frames will prevail over facts. Accordingly, the story in this book is not about alternative facts, but about a new technology that will reframe our human mind and consequently reshape our society and our common future.

So what is the metanarrative for the twenty-first century that will allow us to regain a selection advantage? This narrative will need to account for the human condition, with all its uncertainties, multiple timelines and unknown unknowns, as well as the transitory character of all existing solutions, which are failure-tolerant and always open to revision. It will also need to highlight the potential we have as humans rather than the problems and the risks—the opportunities rather than the obstacles. If we have the right narrative, one that allows us to better understand the impact of AI on our minds, our society and our common future, it can provide a powerful tool to tackle the major challenges we are facing in the twenty-first century. Every technology—whether it be the needle or fire, the steam engine or a bread knife—is neutral in and of itself; it is we humans who determine whether it will be beneficial or dangerous for us. The same is true of the rapid advances in big data, AI and deep learning. And just as every technology has its roots in nature, AI algorithms also depend on natural laws, materials and properties.

I define culture as a set of collectively shared habits, customs, beliefs and norms characteristic of, and approved by, a society at a certain period in time, which allows its members to better communicate with each other and express their identity.[2] Language, arts, technology, formal institutions and science are some examples. The term 'third culture' refers to a subset of cultural agreements concerning how we do research, the impact of technology and the ways we gain information and knowledge about the world and achieve wisdom at the end. It is characterised by the unprecedented advances in science and technology we are currently witnessing. This third culture transcends the established humanities and sciences and is mainly determined by the insights and possibilities that arise out of the new technologies that are being developed. It will not only revolutionise findings in the humanities and sciences, but contribute to a new way of thinking, acting and making decisions and to a new era that we might call a second Renaissance.

The first Renaissance (1400–1600) was characterised by a critical reception of the ancient Greek and Arabic tradition. This period, on the one hand, saw an unprecedented blossoming of human creativity and prosperity, but, on the other, led to further fragmentation of our knowledge.[3] The second, human-centred Renaissance

[2]Wikipedia (2023a).

[3]The Western human-centred approach does have its advantages: it acknowledges humans' unique ability to reflect upon, question and revise their agendas, dogmas and worldviews, and to constantly correct and recorrect their path in a way that is disruptive and failure-tolerant. This approach has allowed humans to overcome Malthus cycles, brought about a revolution in education, science, the arts, architecture, technology, music and crafts, and transformed government constitutions, trade and politics, while at the same time differentiating and dissociating our knowledge about ourselves

will not be a repeat of the first, but will rather seek to integrate our fractured knowledge and wisdom about the world and contribute to a larger, more holistic consciousness than any previous human era. It will also involve critical reception of and dialogue with the Eastern traditions of Taoism and Buddhism.

Current Western thinking seems less prepared than these Eastern traditions to fully grasp the challenges and developments we are facing in the twenty-first century, since we consider our minds and selves to be, firstly, separate from nature and, secondly, material things located in the brain or constituting some ultimate substance. This Western worldview tends to divide up an otherwise connected reality. At a societal level, we thus see entities such as states, communities and corporations; on an individual level, we see isolated egos with singular, utility-maximising behaviours. In the classical Greek model, there is a ruler who rules the world from the outside, an Alpha and Omega and a primary cause that precedes every existence. But today our mind and consciousness is more like a self-organising network, a dynamic process, a web without a weaver, which is constantly changing and has no ultimate cause or creator.[4]

Eastern thinking seems better equipped to understand and process what our mind and consciousness are, and so can better comprehend the foundations on which our society is built and better predict our common future and the impact of new technologies. It sees the world as interrelated at both a societal and an individual level. Everything is connected to everything, everywhere and at all times, and this interconnectedness is not random or chaotic.[5] It does not happen by chance, but is structured around various complementary pairs, whereby chaos and creation, rules and randomness, silence and sound, fullness and emptiness, humility and mastery, irregularity and proportionality are intertwined.[6]

---

and the world around us. But Western universalism has lost its superiority, not only in this general and philosophical sense, but also in a very practical, social, moral and political one. This is especially true when it comes to understanding and explaining public affairs, the human mind and the impact of new technologies.

[4]From a Buddhist perspective, the entire world is an illusion (Maya), in which we are constantly identifying ourselves with objects, desires and ideas, simultaneously creating transitory successful adaptations and harmful deceptions that ultimately cause suffering. These multiple mental identifications eventually create the narratives we use to explain the world around and within us. If we were able to fully disidentify from the world, we would overcome all illusions and would finally see reality as it is: mental states such as 'full emptiness', 'oneness', 'nirvana' and 'the one taste' bring us as close as possible to reality as it truly is, beyond any distorting conceptualisations. It may sound paradoxical, but the new technologies we explore in this book share the property of allowing us to both disidentify from and more fully engage with the world.

[5]The interconnectedness we are now experiencing in the Anthropocene era is similar to the 'ecology of mind' first described in detail by Bateson (1972). It is only through interconnectedness that there can be any meaning. Isolation and abstraction are a universal impossibility, since everything is interconnected with everything else. Strictly speaking, we could study anything through the lens of any discipline.

[6]Western traditions have similar sacred geometric proportions (such as the 'golden rule'), which establish an intrinsic link between beauty, proportion and goodness. In Greek philosophy we find the expression *kalos kagathos*, which means 'beautiful and good'. It suggests that if we want to do

Another lesson the West can learn from the East is the importance of proportion (i.e. balance and harmony). Identifying the right proportions in things can reveal the natural patterns that we use to create and understand everything in nature and society. From a historical perspective, the concern with identifying right proportions is older than any type of analytical or critical thinking. Identifying proportionality enables us to move from a merely analytical, linear, siloed, divided and dualistic worldview to one that incorporates and pursues wholeness, oneness and unity. The purpose of proportion is to reconcile or resolve polarity into some kind of unity, wholeness or greater being. The well-known yin–yang symbol represents this kind of proportion. It is linked to the cyclicality of coming and going, appearing and disappearing. The more aware we become of this cyclicality, the more balance can be achieved. And each yin (the passive, nurturing, female principle of the universe, characterised as sustaining and associated with the earth, dark and cold) also contains some yang (the active male principle of the universe, characterised as creative and associated with heaven, heat and light) and vice versa. If we were to rebalance our thinking about politics, economics, science and technology, we would tap into a deeper understanding of the shadows we cast, the voids we ignore and the ignorance we are pursuing.

Another lesson we can learn from the East is how to reconcile such opposites. The fundamental constituents that make up our reality are equal and interdependent and serve to balance each other. They can exist only in relation to one another. Identifying the proper pairs of opposites is not always a straightforward intellectual endeavour. Picking the wrong pairs could lead to the wrong conclusions and have ruinous consequences. Although right/left, female/male, up/down are easy to grasp, there are other forms of opposites where the correct interrelations are more difficult to establish. For example, humility and self-efficacy, freedom and responsibility, coherence and strategy, control and devotion, modesty and generosity, tolerance and discrimination, efficiency and resilience are frequently overlooked yet powerful complementary opposites on an archetypal level that hold particular relevance for understanding any new technology, as we will see. So if we wish to shift to a more Eastern way of thinking, identifying the right pairs is important. I define complementarity as a relation between two elements that are incompatible yet mutually required, that do not cancel each other out but are both needed to describe an event, a thing or a state of affairs. Examples include location and momentum, energy and time, wave and particle, physical and mental, form and content, substance and process.[7]

---

the right thing and make the right decisions, we need to search for and be exposed to the beauty of correct proportions.

[7] See Bohr (2008), Meyer-Abich (1965), Walach (2010).

Such complementary pairs[8] should not be abandoned but rather contained. Both elements are valuable and meaningful in themselves, but also complement each other. It is rather like the oscillation of a pendulum, or a battery that has both positive and negative poles. Failing to understand these polarities renders us incapable of harnessing the power or 'electricity' of life. Thinking in terms of polarities therefore promises to be a powerful tool that will allow us to see and do more. Eastern thinking provides the ingredients to transcend dualism and materialism, generates oneness and connectedness and opens up a balanced, proportional and sound path for us to follow, which will ultimately lead to a unified world.[9]

This is even more important given that new findings and developments emerging out of AI, deep learning algorithms and the overall process of datafication—such as the Internet of Things (IoT), the global superbrain, quantum computing and robotics—are all operating in a manner more attuned to an Eastern mindset than a Western one. It is therefore not surprising that Taoism and Buddhism in particular promise a deeper understanding of what is going on in the digital technosphere in the twenty-first century and can provide a more substantial answer to the question: *what is it to be human in the twenty-first century?* We will see that humans are always a deficient species, never fully adapted to their environment, who require crutches to survive. The third culture that we will explore in this book can provide unique and unprecedented ways to achieve such adaptation. We will come to realise that evolution is not a ladder with the human species at its top, but rather is made up of infinitely many overlapping circles, with humans playing the role of a marginal string player alongside millions of other species and living beings.

Starting with C. P. Snow's thesis of 'two cultures' (science vs humanities) and Kuhn's theory of scientific revolutions, I will explain how the fourth Industrial Revolution,[10] which we are currently living through, has for the first time in human history provided the ingredients for a 'third culture' and a 'second form of scientific revolution', which is having a significant impact on our brains, our minds and our society as a whole.

---

[8]Humans have developed three ways to identify complementary pairs. The first, perception, involves observing and mimicking nature; a second, analytical thinking, originated in the prehistorical practices of shamans, which were then further systematised in Greek, Arabic and Chinese culture; a third, contemplation, is rooted in the mystical practices that cut across all religions.

[9]Whenever technological progress or innovation occurs in one field, a potential shadow is cast, or a void created, in another. Western thinking excels at celebrating progress and disruptive innovation, but has great difficulty recognising the voids resulting from this progress. For example, the invention of the printing press had a negative impact on oral memory, driving a car has a negative impact on walking and consequently upon our health, digitalisation has had a negative impact on jobs and so on. In short, whenever we progress in one area, we also 'regress' in another. From an Eastern perspective, the divided view of the Western Enlightenment is seen as a form of *avidiya* or ignorance.

[10]The first Industrial Revolution (1820) was characterised by mechanisation, and in particular by the invention of the steam engine; the second (1900) by mass production and electrification; the third (1970) by automation and computer technology; the fourth (2000) by the IoT, AI, deep learning, big data and autonomisation. See Schwab (2017).

This book borrows its title from two predecessors. The first is John Brockman's *The Third Culture* (1996), which brought together cutting-edge contributions from dozens of eminent researchers working in disciplines as varied as cosmology, evolutionary biology, genetics, particle physics, computer science and systems theory. These contributions articulate a *new, scientifically informed way of thinking*. However, Brockman widens the gap between the different cultures, rather than providing a comprehensive argument to bridge or transcend it. As he puts it, 'men of letters' typically 'comment on comments' rather than providing new insights.

A new perspective on the third culture argument was given by John Kagan's *The Three Cultures* (2009), which distinguishes between the sciences, humanities and social sciences. In Kagan's view, the three cultures differ along multiple axes, including their vocabulary, their aesthetic values, their contribution to national interests and the economy and their sources of data and evidence, which in turn lead them to different notions of what is valid, right, coherent and 'objective'. The present book will set out a third kind of argument about the third culture that is almost entirely distinct from these two predecessors. I will show that AI, deep learning and datafication in general have the potential to extend findings in the two traditional cultures of science and the humanities and provide a more integrated, holistic view on ourselves and the world around us. This will also shift our consciousness and our society as a whole. What makes this third culture unique is that, if it is implemented the right way, it can serve as an integrator of information, knowledge and wisdom that will further enhance our collective consciousness and allow us to pursue a better future. I identify the third culture as one of three potential integrators, alongside the *financial and monetary system* and *psycho-technologies capable of altering our minds*: in particular, contemplative practices, the use of psychedelic drugs and adapted educational agendas. These new tools and technologies will provide additional insights that allow us to further refine the concept of *singularity* and the debate on *transhumanism*. I will also offer an alternative answer to the problems of the *black box effect* of AI and the *garbage in, garbage out phenomenon.* And I will provide an adjusted *Turing test* that will help readers to better understand what it means to be human in the twenty-first century and to make sense of this new technology and its contribution to the larger picture of a second Renaissance.

1.2 The Questions behind It All

In a letter from 1610, Galileo complained that the local authorities refused to look through the telescope and acknowledge that the Earth is rotating around the Sun and not the other way round. Galileo argued that every time we advance into new domains of knowledge and possibility, whether through inventions or discoveries, humans need to use technology and adopt a new mindset in order to see better and gain a deeper understanding of the world around and within us. If we had refused to look through Galileo's telescope, we would still think that the Earth is the centre of the universe and that the Sun rotates around it. Or consider Plato's famous parable of the cave: humans are chained up in a cave, watching shadows that are cast on the wall by objects moving behind them. The shadows are misconstrued as reality itself, but in fact are merely a distortion of it. Reasoning—metaphorically represented by leaving the cave or at least becoming aware of the limitations of our perspective—can allow us to overcome this illusion and better understand the world around and within us. And technology is one powerful tool to help us do so.

In 1712, barely a hundred years after Galileo's letter, Thomas Newcomen invented the steam engine and the first Industrial Revolution began. Nowadays, we know that any technology that is invented can have either good or bad effects, can be either beneficial or harmful, depending on how people use it. Under the regime of the fossil energy age, humans have been able to triple life expectancy, reduce child mortality by a factor of ten and poverty by over 90% and create unprecedented wealth and prosperity far beyond the Malthus cycles that determined human life on this planet for centuries if not millennia. But at the same time, the technology that began with the steam engine has created massive social and ecological externalities that the planet and the people of the twenty-first century are suffering from. Species loss, global warming, water stress, land degradation and wars are just the most obvious examples.

The situation in the twenty-first century could not be more similar to the one Galileo bemoaned. Just imagine if we refused to look at the findings that have emerged out of deep learning, datafication or AI. Might that mean we risk overlooking that the world is fully interconnected?[11] Might we fail to understand that this kind of technology can reveal galaxies we would never imagine even existed? Or can speed up computing to analyse nanoparticles humans could never comprehend with their native minds even if they lived 500,000 years? Or can tap into literature from the entirety of human history in less than a second in order to answer a question? And what does it mean, for good and for ill, to have all these potential technologies at hand and to be human in the twenty-first century?

Every technology is ambivalent. Take a bread knife, for instance: we can use it to cut bread, or we can use it to kill someone. This is true of the Haber–Bosch process, nuclear reactions and DNA coding, and it is true of the new technologies emerging in the twenty-first century, such as AI, big data, nanotechnology, robotics, cryptocurrencies and blockchain algorithms. We as humans decide how much use we make of each discovery and invention in order to create the society we want to live in. Technology and research follow natural and physical laws. But their implementation does not: it follows social agreements and contracts, approvals and

[11] One way to look at this is as follows: the amount of water and air on this planet has remained constant throughout history (Berner and Berner 2012). Every time we breathe in—and each of the eight billion human beings on the planet does so about 17,000 times a day—we breathe in the same air as all previous generations. The same goes for every glass of water we drink: we are drinking the same water as every previous generation. We have always been connected, and in the future we will be even more connected. The difference from the past is that we can now measure, scientifically evaluate and influence the degree of connection. See Ford (2016), Utke (1998).

falsifications, majority votes, best guesses, opinions and hypes. And science and technology themselves progress unevenly. Some fields develop faster than others, constantly producing transitory findings that are always open to revision.

Viewed from this perspective, the question of what society we want to live in always comes first, and the question of what technology we use second. We could conclude that anyone who claims that technology alone can solve all problems either does not know what technology can do or what the real problems are, or perhaps is ignorant on both fronts. But determining the impact this new technology will have on what it is to be human in the twenty-first century is tricky, because we humans evolve over time, as do the societies we live in and the technologies we invent.

The main task of any science is not to eliminate uncertainty, since uncertainty is part of our reality, but rather to solve specific problems. Doing so does not cause uncertainty to disappear, but opens up space for new questions and problems that require new solutions. This process is never-ending. In most cases throughout history where science has offered fundamental insights, this has not come out of a rational, analytical process of linear logic, but rather has depended on irrationality and intuition. This dark side of science has been the major source of most human discoveries, for instance those of Heisenberg, Einstein, Mendel and Darwin.[12] And anyone who has ever attended a traditional academic conference can confirm that such gatherings never create one big new idea, but have killed a lot of them. If Einstein is right that we cannot solve the problems of today with the tools and measures that caused them, then we will need to think outside the box. As the Nobel Prize winner and physicist Max Delbrück puts it, ‘When you do science, you potentially change the world much more than Caesar or any of the great military or political figures ever did, and you can sit very quietly in a corner and do that.’[13]

1.3 Respecting Boundaries

One of the major limitations of our brains and minds is that we lack the natural ability to perceive exponential patterns of growth. Doing so requires additional intellectual effort. The results of exponential growth will often be far greater than most people would intuitively expect. If a tree grew by 2.7 mm each year, then after just twenty years it would be 485 km tall. If an economy grew by 10% per year, its baseline would double every 7.2 years. And doubling one cent every day would leave us with 5.3 million dollars after just thirty days. Humans have a hard time anticipating such curves and making appropriate decisions in response. And we can observe that exponential growth is happening in almost all domains. Moreover, the

[12] See Fischer (2014, 2015).

[13] See Delbrück (1978).

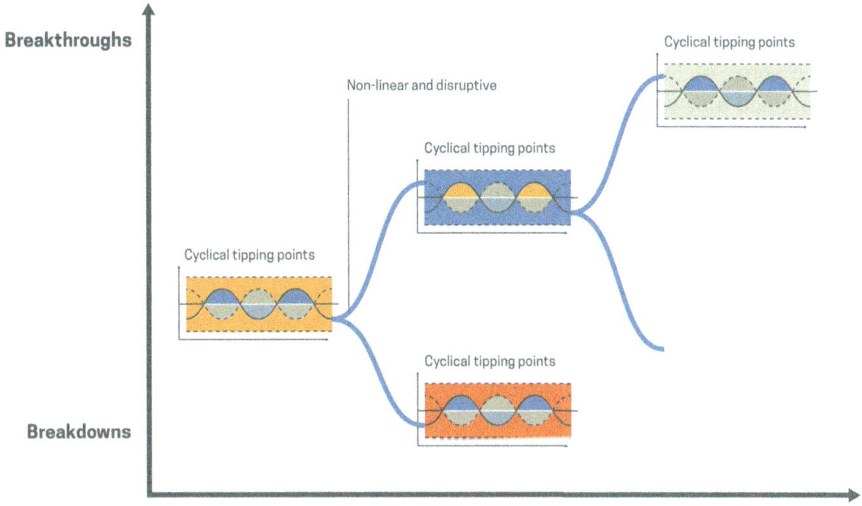

Fig. 1.1 Different stages of societal aggregation: horizontal and vertical evolution

exponential growth itself is growing exponentially,[14] leading to a transformation so rapid it is almost vertical, and which is changing our entire state of societal aggregation. The situation is analogous to water, which can be solid, liquid or gaseous. Each time water changes its state of aggregation, the molecules are rearranged in a different way, rather than new ones being added to it. The same is true of our societal states. Rather than adding new components, we instead need to rearrange the existing components in a more sensible way. Figure 1.1 illustrates this idea. In a first-tier evolution, we select by trial and error, selection and mutation, which is horizontal and cyclical. In a second-tier evolution, we change in a non-linear, disruptive and vertical manner. And the development can occur in both directions: up and down, breakthroughs and breakdowns, to the benefit or the detriment of humankind. Any vertical transformation with super-exponential growth is a challenge, but it also creates super-exponential opportunities and potential to overcome constraints and limits.

Most thinking takes place within a specific conceptual framework, and most (if not all) scientific discoveries emerge out of this rule-based conceptual thinking being challenged and overcome. It is when this happens that irregularities, ambiguities, anomalies and paradoxes are resolved and new connections and insights become visible. The mathematician William Byers calls this mental state 'deep thinking':[15] oppositions and irregularities can be integrated, making complementarities, fractal connections, creativity and new learning possible. If we replace our outdated technology or our government but our thinking remains the same, the new

[14] Kurzweil (2005) claims that there is a 'law of accelerating returns', such that the process of change is itself exponential.
[15] Byers (2014).

technology or new government will be just like the old one. A shift in consciousness towards more mindfulness, grace, courage and humility would allow us to regroup, resolve some of these ambiguities and develop a new paradigm, a new way of thinking and a new way of dealing with the challenges ahead.

Looking back at history, we can observe that the first Enlightenment (1685–1815) and the first Renaissance (1400–1600) derived their momentum and impact from differentiating and separating the world. Religious beliefs and secular life, economy and ecology, state and market, collective and individual are some examples. But we now appear to be entering a second Enlightenment and a second Renaissance, whose primary impulse is not to further compartmentalise and separate the world around us, but rather to integrate, incorporate and embrace it at a higher level of consciousness. Some scholars call this process the 'great convergence', running in parallel to the 'great acceleration' in which humans are now sitting in the driving seat and determining the course of our planet. Politics, science, economics, religion and reasoning in general are being reconciled into a greater whole. One of the most powerful integrators in this process is the new technologies that are emerging right now.[16] The 'great convergence' relies on the human capacity for creativity and understanding, which enables us to synthesise things rather than further separating them. In short, it is about integrating the Good, the Beautiful and the True rather than emphasising the differences between them.

Box 1.1 The emergence of new technologies
New technologies, inventions and discoveries are always ambivalent. Most research findings are not the result of a rational process, but rather emerge from science's intuitive, irrational 'dark side'.

By 1820, most of the ingredients for change were already in place, with one exception. Humans had invented the wheel and the printing press and learned how to make fire. There were already nation states, a banking system and a tax system. Mathematics, astronomy, religion, music and art as well as knowledge of human anatomy were already established. Copper, iron, wheat, meat, fruit, vegetables, bread and butter were also available. Most of the elements of daily life as we know it today were in place. Despite this, people's living conditions had not changed significantly for centuries, if not millennia. Human life remained the same from birth to death, and societies as a whole evolved according to what is known as the Malthus cycle: economic growth was driven solely by demographic factors. But around 1820, some thirty years—just a single generation—after the French and American revolutions, something amazing happened that set in motion an entirely new process, unlike anything seen before in human history: the social empowerment of the

[16]There are three additional integrators. On an individual and interior level, our spiritual practices; on a collective level, our commonly shared values; and on a systemic outer level, the architecture of the financial system. All three have the potential to fundamentally change our minds, our consciousness and the future course of humankind for good or for ill. See Brunnhuber (2021b, 2023a).

individual to use their critical mind, creativity and new forms of social cooperation. This human-centric mindset changed everything. And we are currently witnessing another comparable change in our mindset, for which most (if not all) of the ingredients are already in place.

In this book, I explore the fundamental points of contact between AI, big data and deep learning, on the one hand, and our human consciousness, on the other. I look at the role of traditional science and the humanities, and the paradigm shift that is resulting from new findings and developments in computing. Against the backdrop of this ongoing debate, I explore links to the Turing test, transhumanism and the concept of singularity and discuss how computer technology can help us understand what it means to be human in the twenty-first century. As we will see, this involves different forms of learning and acquiring knowledge. I conclude that we may be witnessing the dawn of a 'third culture' that could potentially mark the beginning of a new integral wisdom.

The first Axial Age (2,500 years ago), the first Renaissance (1400–1600) and the first Enlightenment (1750–1820) were all about increasing differentiation between the outer and inner worlds and further compartmentalisation and specialisation of our knowledge. The second Enlightenment or second Renaissance that we are now witnessing primarily involves integrating the results of this differentiation process. In this great convergence, politics, science, religion, thinking and action come together and are reconciled. The three most powerful integrators in this process are information technology, spiritual practices and the financial sector. Figure 1.2 illustrates this along a historical timeline.

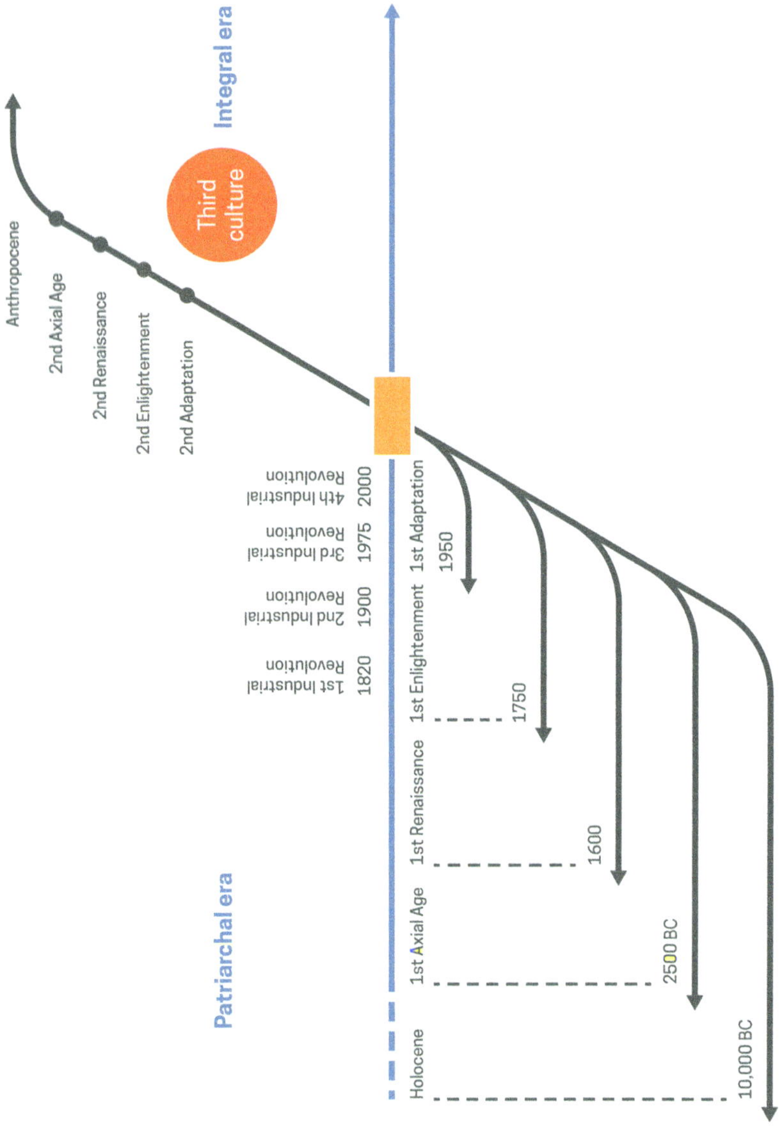

Fig. 1.2 The dawn of a new era: towards integral consciousness

Chapter 2
Finding Potential Integrators

The modern age began by disintegrating, deconstructing and questioning our reality. Rather than taking that reality for granted, people explored alternative ways to look at the world and improve our quality of life. They began studying the laws of nature and the rules of government, and invented the printing press, the steam engine and antibiotics. In the twenty-first century, humans will start integrating all the knowledge and information we have gathered over the centuries. But integrating knowledge is a fiendishly difficult problem. It does not come for free or happen automatically. Integrating fragmented information or isolated opposites often requires a third party or entity. For example, if we want to see the world in three dimensions, we need the left and the right eye, and we need the orbital cortex to integrate the two into a 3D picture of our reality. An integrator must be able to transcend differences in political or ideological agendas, to increase our awareness and perception and/or to reconcile different empirical findings and information into a greater whole. There are three candidates that could potentially serve as integrators.

(a) *A reformed financial sector:* In order to integrate the allocative power of a free market system with different forms of state intervention and to reconcile the different political agendas of autocracies and open societies, the involvement of a third agent may be needed. The most prominent candidates are central banks and regulators. I will show that upgrading the mandate for regulators can have a significant impact and allow us to fund the gap in common public goods, to unlock and de-risk trillions of dollars of private-sector capital and to overcome the oppositions between different political agendas.
(b) *Altered mindsets:* A second integrator of fragmented knowledge and worldviews can be found in the findings of cognitive science and ancient wisdom traditions. Humans are able to alter their mindset using a set of contemplative tools, educational agendas and spiritual practices (including the use of psychedelics). All three methods have, independently of the others, the capacity to alter our minds, but using two or three in concert can increase their effect on our

S. Brunnhuber, *The Third Culture*, Sustainable Finance,
https://doi.org/10.1007/978-3-031-48113-0_2

consciousness and allow us to achieve a metacognitive state that opens us up to a greater wholeness.

(c) *New technologies and the third culture:* A third potential integrator comprises new technologies that are able to integrate and enhance findings in the sciences and humanities and provide the platform for a 'third culture' that will further enhance and integrate our knowledge about ourselves and the world around us. Whereas the first two integrators are explored in the next few sections, the third integrator is the topic of the remainder of the book.

2.1 A Reformed Financial Sector

Our financial market is one of the few international institutions that most, if not all, actors operate within. Despite their different political agendas, every country—whether it is an open society or an autocracy, a developed economy or a failed state—operates within the existing monetary system. Even terrorists, the black market, fraud, illicit transactions and corruption depend in some way or another on a functional global financial market. In this general sense, the international global capital market, its associated institutions (IMF, WB, central banks) and its monetary policies determine the rules of the game for fiscal policy, real economic activities, non-profit commitments and household spending. They act like an attractor for the good and the bad. For as long as we overlook the crucial role of the monetary system, we will fail to understand its relevance and potential integrating function.

Traditionally, money has been excluded from the equation that can be represented as a triangle between the real economy, the social world and the environment. Doing so paints a misleading picture: the monetary system has always been there, acting like an attractor at the core of our society. Figure 2.1 illustrates this.

Fig. 2.1 The money system serves as an attractor and integrator

However, money is not a thing or a natural law, but rather a convention, a social mechanism, a club rule and a set of legal codes that we can change as we see fit. Its current configuration runs counter to the goal of a sustainable future and prevents us from integrating the market systems of the real economy, its systemic social and ecological externalities and the dynamics of the financial market. There are six core elements to the current configuration of the monetary system, which are outlined in Table 2.1 below.

Table 2.1 Money is not neutral—six elements of the current system that are damaging our future

Pro-cyclical amplification of boom and bust cycles	Banks provide and withhold credit lines according to the requirements of the real economy and amplify the cycles, instead of counteracting them
Short-term perspective	A discounted cash flow enforces short-term decision-making
Compulsory growth	A compound interest rate forces states, companies and households to grow to pay back their debts
Destruction of social capital	Instead of encouraging trust, solidarity and cooperation, fear, greed and parasitic competitiveness are enforced
Widening income/wealth inequalities	Current incentives support financial assets rather than real economic needs, further increasing the income and wealth gap
Multiple rebounds	Efficiency gains are cancelled out by increased consumption that further hinders progress towards a sustainable pathway

If we had a different monetary system that respects the achievements of the existing financial architecture but remedies its flaws, we might have a powerful tool to integrate a dissociated financial market and create a healthier, wealthier planet. *Traditionally*, we mainly redistribute money from the private sector in order to fund, manage and hedge public goods and global commons.[1] However, this 'end-of-pipe' approach is slow, administratively demanding, small in scale and insufficiently targeted to meet the challenges we are facing in the Anthropocene era. Figure 2.2 illustrates this:

[1] The argument of 'effective altruism' is a prominent example. A hedge fund manager could pledge their income to a deworming campaign and do far more good than if they quit their job and became an organic farmer. But this approach operates within the existing financial system and assumes that it functions properly, when the reality is that it is flawed from the ground up. In short, we need to upgrade the system to meet the requirements of the twenty-first century, instead of merely working with or around the existing system.

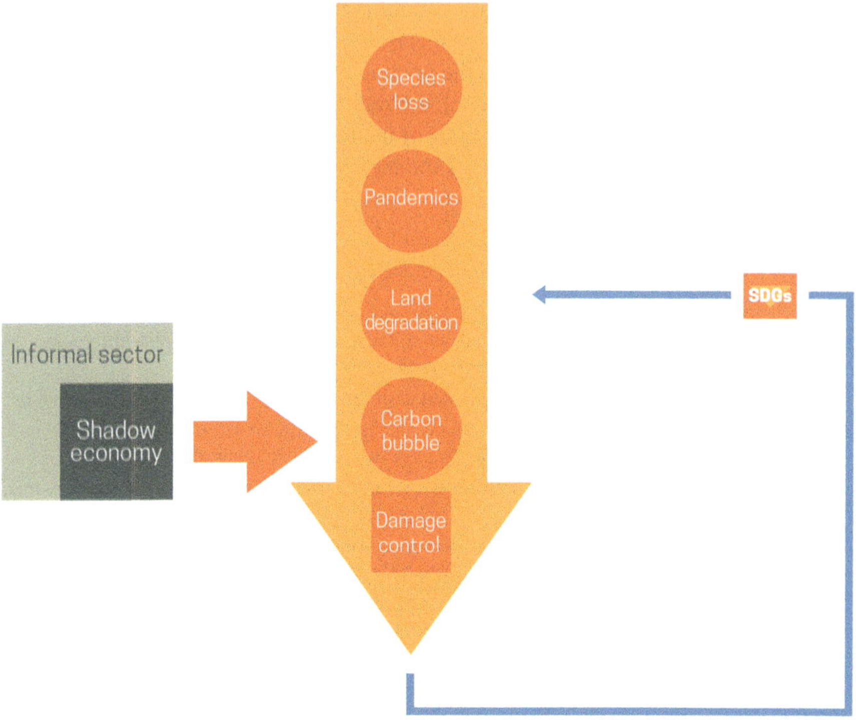

Fig. 2.2 The traditional way to do it—end-of-pipe, redistributive approach

The large orange arrow represents the entire value chain. Any time we produce something, we are contributing to species loss, climate change and land degradation, which in turn incur additional costs as we must manage the resulting damage. 79% of the value chain is still dependent on fossil energy.[2] At the same time, the entire economy is affected by the shadow economy (grey box), which pulls all economic activities in the wrong direction. The small blue arrow represents transfer payments (philanthropy, taxation, official development aid (ODA) or green impact investment). In this standard approach, we first generate social and ecological externalities, then create a compensation mechanism and finally fund global commons.

[2] Assuming a 3% global growth rate, the total conversion rate (TCR) from fossil to green energy would need to be roughly 5% per year to override the growth dynamic. Any time we build a wind turbine or install a solar panel, we still generate income and revenue that is 79% dependent on fossil energy. This is one reason why we need to upgrade (parallelise) the currency system to incentivise green investments and generate multiple positive second-round effects to help bring about a sustainable future (Brunnhuber 2021b, 2023a).

The UN-SDGs as a Case Study

In New York in 2015, world leaders signed up to a future roadmap with seventeen Sustainable Development Goals (SDGs) intended to benefit people, planet, prosperity, peace and partnership. Most of these SDGs focus on common goods such as clean air, universal access to healthcare, education and biodiversity. These goods are not exclusive and should be accessible to and enjoyed by everyone. All the goals have enough scientific evidence, technological know-how and political consensus behind them to be achieved, and they apply to the entire planet. But meeting the goals will be expensive, requiring approximately 5 trillion USD/year over the next fifteen to twenty years to finance.[3] Our global gross domestic product (GDP), which includes all goods and services, is approximately 100 trillion USD/year. The conventional way to finance social and ecological projects globally has been by redistributing the money remaining at the end of this pipeline. Historically, the world community has committed to spend 0.7% of global GDP—roughly 700 billion USD/year—to finance common goods. Other than the Scandinavian countries, the vast majority of the world has never attained this 0.7%. But even if all countries were to meet that commitment, it would realistically not be enough to finance our future. Approximately eight to ten times as much funding—equivalent to 5 trillion USD out of the 100 trillion USD global GDP—is required to meet the social and environmental challenges we face. Withdrawing 5 trillion USD from the ongoing economic process, even in a gradual manner, would lead to a global recession. Withdrawing this amount of money would reduce the capacity of the private sector to transform itself in line with the UN-SDGs. In fact, it is impossible to finance our future solely through monetary redistribution. In addition, the stability of the financial system itself is an impediment to sustainable financing. Over the last forty years, the financial system has become more unstable, with over 425 banking, monetary or currency crises; and every consecutive crisis leaves us with a higher debt load and greater expenses, amounting to more than 10% of GDP. Because of this, the world community puts great effort into repairing, stabilising and refunding the monetary system to maintain the status quo. This limitation in our financial system hinders any technological or political attempts to make the world a better place. Is there a different way to finance our future?

Traditionally, there are at least five steps we can take to fund, hedge and manage our commons. (1) Philanthropic pledges, ODA, grants and bonds; (2) private equity, including venture capital, seed investments, early-stage investments, SME-transition

[3] One prominent example is the funding gap for the 160 million micro, small and medium-sized enterprises (MSMEs), which amounts to over 5.2 trillion USD globally. Three-quarters of MSMEs do not even have access to bank loans. This is a sign of capital market inefficiency. High interest rates, complex administrative procedures and a lack of collateral mean millions of firms cannot access adequate liquidity. *Open banking*, where the financial institution has direct access to the balance sheet (data in motion principle), can reduce costs, increase trust and liability, generate bottom-up alternative data and allow secondary debt market scaling (mortgage-backed securities). See People-Centered Internet (2023).

funding and large-scale institutional investments; (3) multiple taxation schemes;[4] (4) private–public partnerships (PPPs), including synthetic asset-backed securities (ABSs) and state guarantees; (5) regulatory efforts, including ESG taxonomies, prioritising investments according to their public return on investment (ROI)[5] and multiple financial disclosure directories. However, if we take all these measures together, there will be a financial funding gap, which leaves us unprepared to fund our commons, to unlock and de-risk private capital and to manage social and ecological externalities. There is increasing evidence that an adjusted monetary aggregate is required to provide adequate and conditioned liquidity so as to finance future shocks, hedge associated systemic risks, prevent bank runs and enable a secondary preventive strategy.[6] There are two major facilities, namely CBDCs (central bank digital currencies) and CBCSs (central bank currency swaps), that could play a crucial role in filling the financial gap we are facing.[7] The following sections will look at them in more depth.

Systems Thinking and DeFi vs CeFi

Systems thinking can give us an out-of-the-box solution to generate the funds needed to finance global common goods *and* to de-risk and mobilise trillions of private-sector liquidity at the same time: (a) central banks would be given an adjusted monetary mandate to create and issue the required liquidity using blockchain technologies. Alternatively, (b) properly regulated corporate initiatives

---

[4]This includes VAT, a harmonised international corporate tax, an enlarged tax base, reduced tax expenditures and an earmarked 'sin' tax. However, any taxation scheme will have multiple downsides: it will require international agreements, it will incur high administrative costs and its steering capacity will be limited due to its regressive nature. Moreover, companies do not have the money to fund their own transition. In an optimistic scenario, global taxation schemes could generate around 250–350 billion USD per year. Taxation is part of the solution, but cannot be the entirety of the financial transition plan.

[5]If we start prioritising the UN Agenda for Sustainable Development, using an ROI analysis and taking less spectacular but highly efficient and highly preventive measures, we can do more good than by simply providing 175 billion USD in ODA per year. Such measures include investing in education (ROI 1:30), maternity and postnatal health (ROI 1:87), anti-malaria campaigns (ROI 1: 48), improved nutrition (ROI 1:33), child vaccination (ROI 1:48) and skilled migration (ROI 1:20); stimulating trade and specialisation (ROI 1:7 for OECD countries and 1:99 for LDCs); and introducing a sin tax on nicotine, sugar and/or alcohol (ROI 1:23). These latter measures could save 4.2 million lives per year and an investment of 35 billion USD/year would have a social benefit of 1.7 trillion USD. See Lomborg (2023).

[6]Primary prevention refers to preventing harm or damage in the first place; secondary prevention to addressing the future costs of harm or damage that has already occurred; tertiary prevention to management of a chronic state. With regard to the challenges of the Anthropocene (climate change, pandemics, species loss, etc.), we are confronted with a secondary preventive scenario: we have caused the damage already. Now we have to manage the potential future costs associated with that damage. An extended monetary aggregate that gives priority to fiscal policy can serve this purpose.

[7]See Atlantic Council (2023).

(cryptocurrencies) or complementary communal currencies (such as local exchange trading systems (LETSs) or regional money) would receive a mandate to issue additional liquidity. These funds would be earmarked and used exclusively to finance SDG-related projects. This electronic liquidity would run through different monetary channels than those of the conventional system. We would then have a supplementary currency operating in parallel to the conventional monetary system, generating the 5 trillion USD equivalent needed annually for the next twenty years. Research on optional parallel currency systems has shown dozens of positive effects. For example, this new technology could be used to create and channel targeted financial liquidity to millions of people in Africa through their mobile phone network. In India, the existing microcredit banking system could be used to transfer additional liquidity to millions of its citizens. Any dollar spent and invested through these green, parallel channels would have the potential to reduce or even eliminate absolute poverty globally within less than one year. The electronic format would prevent corruption and fraud, as each transaction would be transparent and public. Once the currency became an eligible means to pay taxes, government agencies would have additional liquidity to rebuild public infrastructure such as nurseries, parks, hospitals and libraries. And the millions of NGOs globally would finally receive the funding they need to do their jobs properly. This targeted additional liquidity would enhance education and provide access to universal healthcare that would otherwise never materialise. It would reduce resource depletion and clean up the air, preventing negative effects on our planet and on public health. We would eventually unlock the untapped potential of millions of unemployed people by creating new jobs, which would in turn unlock the creativity of billions more people.

Box 2.1 The web without a weaver paradox: DeFi and CeFi

The term 'decentralised finance' (DeFi) refers to business models that allow transactions without intermediaries such as banks. A token, created, for instance, by an initial coin offering (ICO) and linked to a blockchain-based smart contract, provides the additional liquidity needed for the transaction. At present, DeFi business models have a volume of 42 billion TVL (total value locked) and remain a niche product. Fully decentralised finance is a myth. Creating and maintaining any DeFi models will always require some sort of hierarchy. All DeFi models to date have failed because they (often surreptitiously) involved some sort of centralised finance (CeFi). The crypto stock exchange FTX, the stablecoin TerraUSD and the crypto bank Celsius are recent examples.[8]

Instead of further decentralising the financial and monetary system, we should start by upgrading it.[9] And instead of looking for the smallest common denominator

[8] Meyer, Welpe and Sander (2022), EUBOF (2022).

[9] Brunnhuber (2021b, 2023a).

between state and market, economy and ecology, we should introduce a third party, namely regulators and central banks. This idea of triangulation, adapted to digitalisation, is one of the cornerstones of any integrator.[10] If we take this approach one step further, we will end up with a more competitive marketplace and a stronger, more resilient state at the same time. Figure 2.3 illustrates this:

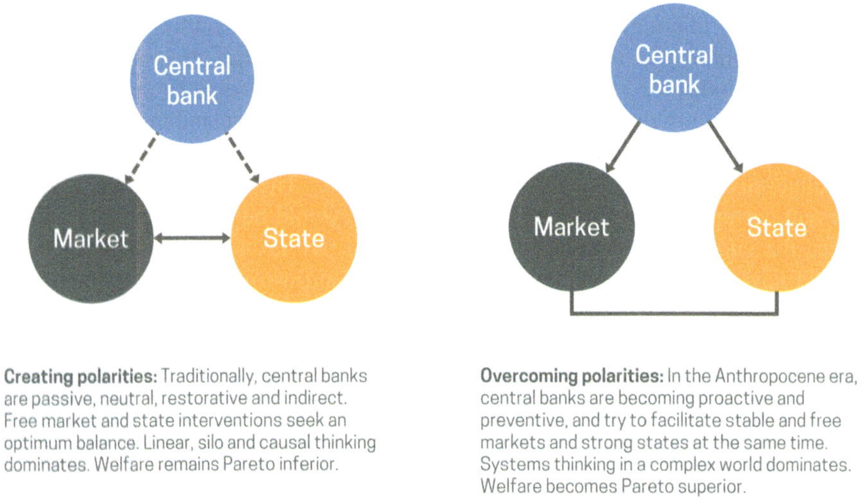

Fig. 2.3 Overcoming polarities: triangulating the system

The Ultimate Step: Central Bank Currency Swaps (CBCSs)

Central bank currency swap lines are one of the hidden monetary and financial champions that could allow us to meet the funding and hedging needs of a global commons. Technically speaking, a currency swap line is a political agreement between two countries' central banks to exchange their domestic currencies with each other. Swap lines can be limited or unlimited, bilateral or unilateral. For example, the Eurozone has established unlimited swap lines with the US dollar,

[10]The Nash equilibrium refers to a situation where, given a certain set of rules, opposing agents reach a position in which they are no longer able to collaborate without harming their own position. In order to overcome this lock-in effect, the agents must change the rules of the game or introduce a third party accepted by both agents. Due to multiple lock-in effects, we currently find ourselves in a Nash equilibrium on a global scale: North versus South, state versus market, environment versus economy, and so on. In order to transcend these oppositions, we need to introduce a third party that fundamentally changes the rules of the game and maximises the outcomes for all agents involved. Regulators and central banks could play that role. For Nash's original account of the eponymous equilibrium, see Nash (1950) or Brunnhuber (2021b).

allowing it to settle the face value of currencies.[11] Theoretically, any country with the sovereignty to print its own money can do so without limit. If a country is indebted in a foreign currency and is facing imported inflation, bilateral or multilateral currency swap lines can be used to tackle the challenge.[12] To see precisely how this might work, let us take the Amazon rainforest as a case study.

Box 2.2 A case study: rescuing the Amazon with a CBCS

Brazil has debts in foreign currencies (mainly USD) equal to 30% of its GDP, or over 600 billion USD annually, and is facing an inflation rate of over 8%. At the same time, Brazil is the owner of the Amazon rainforest, which is a critical tipping point for the global climate (global common good). Currently, Brazil is deforesting the Amazon at a rate equivalent to 2,000 football fields per day. The land is used for further resource extraction and palm oil/soya production. This creates thousands of domestic jobs and meets the demand of the Global North. However, deforesting the Amazon comes at the cost of enormous negative spillovers that harm all humanity. It has been calculated that the Amazon has a face value of about 250 billion USD.[13] Economically speaking, Brazil cannot afford *not* to burn down the Amazon. However, a central bank currency swap (CBCS) line could fundamentally alter the playing field. If the IMF, Fed and ECB were to provide Brazil with a conditioned currency swap line allowing it to convert a portion of its own currency (the real) into USD or euros, Brazil would be able to pay back its external debts, reduce imported inflation and convert the deforestation industry into a green industry in which rangers and indigenous peoples are paid to preserve the Amazon instead of burning it down. This would further provide a positive spillover for the Global North, as the global temperature would be stabilised, the tipping point would be avoided and the costs of disaster management due to wildfires, floods, heatwaves, etc., equal to about 5 to 7% of GDP in OECD countries, would be reduced over time. A monetary agreement along those lines would be a non-regret approach for all parties involved. The central banks' balance sheets would expand[14] and our palm oil would become more expensive to reflect the social and ecological externalities.

---

[11] The IMF's special drawing rights (SDRs) are a special case. Over 95% of the 450 billion SDRs (as of 2021) are used by OECD countries and MICs.

[12] Access to bilateral swap lines is mainly restricted to OECD countries. 99% of least developed countries (LDCs), 95% of landlocked developing countries (LLDCs) and small island developing states (SIDSs) and 70% of middle-income countries (MICs) have no access to such agreements. See Perks et al. (2021).

[13] Banerjee et al. (2022), Silva et al. (2022).

[14] To be more precise: the global currency market is the largest and most liquid capital market, with around 7 trillion USD equivalent in turnover per day (!), including all assets and facilities. Injecting an additional 250 billion USD equivalent to buy up the Amazon over one to two years will not have an impact on the face value of any major currency.

Green Financial Transition Planning

The aim of a financial transition plan is to provide the liquidity, financial assets and resources necessary to achieve certain targets as efficiently and effectively as possible. The targets themselves are generally set by political and societal consensus. The UN-SDGs are examples of such targets. They require an additional 5 trillion USD of liquidity and financial assets per year to be achieved over the next ten to fifteen years. The 5 trillion USD annually would be the compound result of additional, conditioned liquidity, assets and measures taken to achieve the UN-SDGs within the next two decades. The greater the systemic risks, the greater the role for public bodies and a monetary aggregate. So compiling the fourteen components of this green financial transition plan will be a political decision. Table 2.2 outlines an initial proposal.

Table 2.2 Outline of a green transition plan

	Key principle	Characteristics
1.	Prioritisation	Prioritising based on ROI, KPIs and empirical evidence
2.	Regulation	Taxonomy, procurement (no excess profits, conditioned, mission-based), shadow economy and informal sector, multiple disclosure directives
3.	Reform	Institutional reforms involving the World Bank, IMF, public development banks and the European Investment Bank
4.	Philanthropy	Operating as grants, loans or bonds, often conditional, crowding-out effects
5.	Taxation and subsidies	VAT, earmarked sin/corporation/wealth/international tax, off-shore sites, expanded tax base, tax spending, growth-dependent, administrative overload, reduced steering capacity
6.	Credit lines	On-balance-sheet (credit to bank), off-balance-sheet (special purpose vehicles), MBSs (mortgage-backed securities)
7.	Debt restructuring	Maturity, interest rate, debt-to-X swaps, Paris Club
8.	Private equity	Impact funding, venture capital, start-up (co-)financing, cherry-picking, short-termism, risk-averse, liability
9.	Hedging risks	Credit spread, longevity swaps, foreign currency derivatives, inflation swaps, asset class risk stratification, CDS, interest rate
10.	Private–public partnerships	Asset-backed securities (ABSs), advanced commitment strategy (ACS), public bank endogenous credit creation (PECC), public–private equity share (PPES)
11.	Special drawing rights	Providing additional liquidity to fund/hedge and manage public goods, access for lower and middle-income countries (LICs, MICs)
12.	Central bank digital currencies	Upfront (direct) loading, funding, hedging, managing commons, de-risking private sector involvement, e-wallets
13.	Central bank currency swaps	Conditioned currency converter for LICs and MICs to tackle imported inflation and finance the UN-SDGs
14.	Quantitative easing	Zero coupon perpetual facilities, preventing bank runs, non-defaultable loans (NDLs), operates like a monetary anchor

5 Trill USD/y

What would be the effects on the conventional economy? The annual 5 trillion USD equivalent of added liquidity would not harm the conventional economy. In fact, the opposite would be the case. Corporate and state planning, production and price levels would become more robust and reliable with a longer-term vision. Furthermore, it would stabilise the economic cycle of booms and busts. Despite arguments to the contrary, we need much more financialisation (finance/GDP). However, that financialisation must be designed in a more democratic and humane manner, so as to protect the planet while increasing wealth for two-thirds of the global population. If there is a single most important variable besides technology, governance, behavioural changes and demographics when it comes to changing the world, it is new, digital financial engineering instruments. That would be a real game-changer and could be set in motion in less than six months, if the six largest central banks agreed to create a parallel, optional, complementary currency. Redesigning the financial system would not solve all our problems, but it would make them easier to address. This, or some similar mechanism, is the missing link to achieving better outcomes in terms of people, prosperity, peace, planet and partnerships. If we want to think outside the box and consider an alternative approach, our financial system will be pivotal.[15] Figure 2.4 illustrates the entire monetary upgrade that will be necessary:

[15]For further details, see Brunnhuber (2021b, 2023a) and the WAAS initiative 'The TAO of Finance': https://new.worldacademy.org/tao-of-finance/.

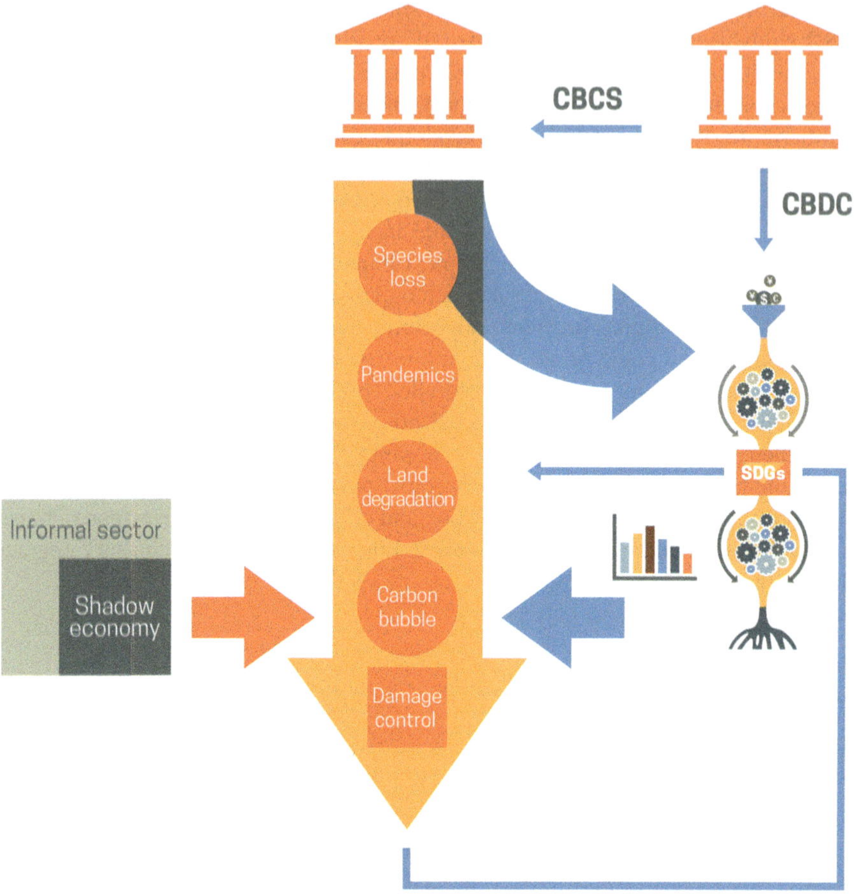

Fig. 2.4 Rethinking finance: CBDCs and CBCSs can provide the liquidity and leverage to fund, manage and hedge our commons. (SDGs: Sustainable Development Goals; EGD: European Green Deal; CBDCs: central bank digital currencies; CBCSs: central bank currency swaps)

Besides the financial sector, which can serve as one integrator, there is a second powerful tool that has the potential to change our world: namely, altering our minds. That is the topic of the next section.

2.2 Altered States of Mind

Most thinking happens within a pregiven conceptual framework, and most (if not all) scientific discoveries occur when this rule-based conceptual thinking is questioned and transcended. Irregularities, ambiguities, anomalies and paradoxes are dissolved and new connections and insights become visible.[16] If we replace outdated technology or our government but our thinking remains the same, the new technology or government will, in effect, be no different from the old one.[17]

A shift in our consciousness towards greater mindfulness, grace, grit and detachment would allow us to regroup, resolve some of these ambiguities and generate a new paradigm, a new way of thinking and a new way of managing problems. The human brain is not only the most adaptive organ but also the most (self-)deceptive. Frames and biases, shadows and echoes shape our reasoning and can be maladaptive and reductive.[18] In order to overcome these biases and flaws, we need to think outside the box. The nine dots puzzle, which is well known in cognitive science, can illustrate this. To solve the puzzle, you have to connect nine dots using four straight lines without lifting your pen.[19] We can only do that if we think outside the box, which in this case means thinking outside the square made up of the nine dots. The puzzle and its solution are shown in Fig. 2.5 below.

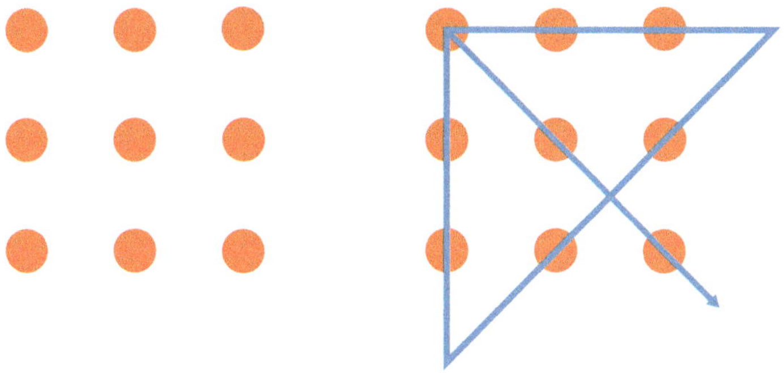

Fig. 2.5 Thinking outside the square box—the nine dots puzzle

Any altered state of mind could potentially open up more integral meaning and morals, wisdom and understanding, realness and connectedness. Enlightenment means overcoming self-deception and alienation. In order to alter our state of consciousness, we have to change not only our frames and way of looking at the world, but also our very selves. About 30–40% of the population have experienced altered states of consciousness in some form or another over the course of their lives.

[16]Byers (2014) calls this state 'deep thinking': opposites and irregularities can be contained, so that complementarities, fractal correlations, creativity and new learning can occur.

[17]One of the most powerful frames is the 'confirmation bias': we favour information that confirms our existing beliefs and values. Some scholars consider it one of the most misleading aspects of human thinking; see Oswald and Grosjean (2004).

[18]It should be noted that when it comes to bullshit and fake news, the truth is irrelevant; the liar knows the truth.

[19]Lung and Dominowski (1985).

In these states, we can learn not only to reason differently, but to do things differently. Grasping different levels of reality can lead us towards greater oneness, wholeness and coherence. Following this path is not a matter of intellectual belief, but rather requires processual, performative and participatory knowledge and engagement in different practices. There are three main ways[20] such a shift can be achieved: through contemplation, through education and (to a limited degree) through psychedelic drugs.

Contemplative Practices

Religion is, on the one hand, the single main cause of suffering, war and social exclusion. But on the other, it is the single most important tool for deliberation and salvation. To better understand this paradox, we must distinguish between the exoteric and esoteric aspects of religion. The exoteric aspect comprises institutional structures, authorised texts and rules. It provides a narrative that interprets the world around us in a certain way, with each religion offering its own perspective. The esoteric aspect, meanwhile, relates to the inner, subjective perspective. It can provide a common ground of timeless truth and offer tools and practices to transform our personal consciousness. These mystical traditions are based on a participatory wisdom that encourages each individual to pursue a specific contemplative practice that goes beyond simply reading a sacred text or following a certain dogma or rule. Purification techniques (e.g. fasting, silence, reduced sensory input, *via negativa*, hermitism), repetitive mantras, mindfulness exercises (e.g. yoga, full-body submersion), martial arts, kōans, rosaries, etc. support and encourage an ongoing process of disidentification. Instead of reading, discussing and interpreting sacred texts and commandments, which teaches people to interpret the world and the self in a certain way and stabilises their worldview, the contemplative practices that exist in all religions encourage people to disengage from their own beliefs and emotions in order to alter their mindset and foster altered mental states that transcend the individual ego.[21] Table 2.3 presents a summary:

[20] Vervaeke (2020), Wilber (2022).

[21] This includes post- and transpersonal mental states such as kindness, humility, grace, reverence and gratitude. Terms for these altered mental states include 'unconditional love' (or unconditional empathy, unconditional forgiveness), 'one taste', 'nirvana', 'absolute emptiness', 'samadhi' and 'inner peace'.

Table 2.3 Differences between the exoteric and esoteric aspects of religion

Religion: exoteric aspect	Spirituality: esoteric aspect
Provides a narrative about the world	Provides psycho-technological practices
Translational—horizontal	Transformative—vertical
In-group experience	Individual experience of oneness
Rules, great books, authorities	Ongoing process of disidentification
Provides stability—hierarchy	Transpersonal mental states
Determined by the past—history	Pulled by the future—attractor

Since both aspects are necessary and common to all religions, they both need to be upgraded and understood in greater depth in order to act as a potential integrator. Whereas the outer, institutional aspect is partly broken, providing us with a false and outdated narrative that is unable to integrate rational thinking and scientific evidence, the esoteric aspect of any religion is intended to guide us and enable us to decode the timeless truth they all share. Figure 2.6 below demonstrates the shift in our mindset away from the ego-state.

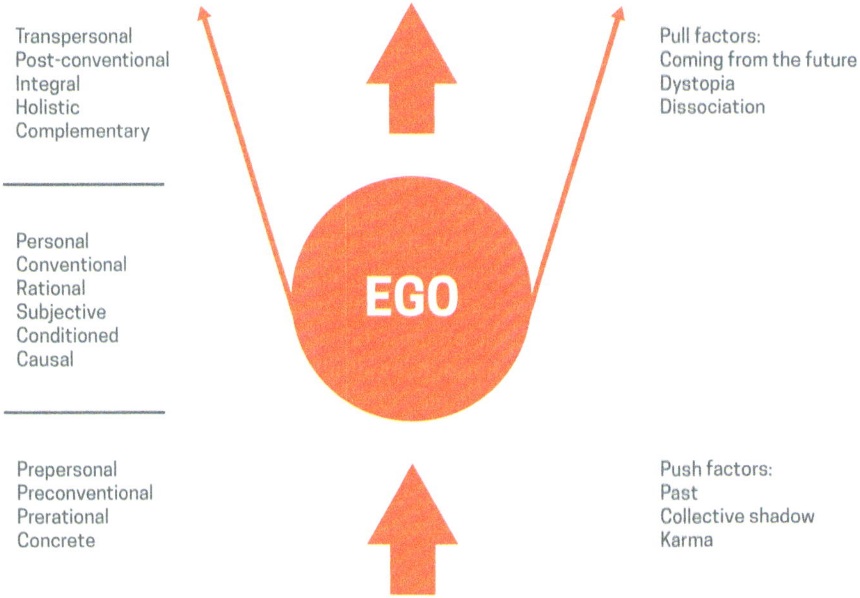

Fig. 2.6 Push and pull factors leading to altered, more integral states of mind

Why a Different Educational Approach is Needed Now More than Ever: An Initial Exploration

Education is always ambivalent. On the one hand, it has to pass on accumulated knowledge and experience to the next generation. On the other, it also needs to prepare them for an unpredictable future and enable them to change the world. That includes skills and knowledge, character formation and the ability to adapt and become resilient in the face of the ongoing changes and asymmetric shocks that characterise the Anthropocene era.

The challenge is that knowledge and information-based education has an average shelf life of five years. In ten years, two-thirds of the companies we are supposed to work for will no longer exist and 40% of our jobs may have been replaced by robots. The current educational system was developed in the nineteenth century and it would take thirty years to replace it.[22] At the same time, educational interventions are one of the few social interventions that have been shown to have a positive causal link to happiness, longevity, economic growth, prosperity, democracy, number of children and a sustainable future,[23] with no significant negative side effects or externalities. No other human intervention is capable of that. The more personalised our education, the better. And, of course, it is crucial that we do not stop learning: any time we intentionally refuse to learn something, or delegate that task, our mind and brain will learn to *refuse to learn* or to *delegate learning*. However, a Western perspective that emphasises disruptive advances and groundbreaking innovations has difficulty identifying and responding to the losses that come with new technological inventions and discoveries. Take the printing press: groundbreaking in terms of educating people and teaching them how to read, but leaving a negative impact on orally transmitted memory. Or take transport: the invention of the automobile improved our ability to travel, but had a negative impact on our health. Digitalisation has increased efficiency, speed and knowledge production, but left us with numerous negative side effects.[24] In short, any time we make progress in one area, it will also bring with it deficits and losses in another. The outsourcing of human qualities and capacities to the digital technosphere is one example. Once we start delegating tasks to a technology, we decondition our skills and lose the capacity to do those things

[22] There is increasing empirical evidence of an input–output fallacy in education. The amount of input (money, teachers, facilities, electronic devices) is only weakly correlated with output (creativity, productivity, well-being). We need to take a very different, far more radical approach to education. See Brunnhuber (2017, 2021a).

[23] See Lutz and Klingholz (2017).

[24] This includes reduced attention span, lack of focus, reduced emotional, social and fine motor skills, a propensity to addiction, reduced development of the prefrontal cortex and the impact of loneliness, particularly in the evolving brain during the first two decades of life. Not to mention the most obvious negative impacts: back problems (due to bad posture) and obesity (due to lack of exercise). See Spitzer (2012, 2019) and the literature he refers to; Spitzer concludes that the higher the investment in IT, the poorer the educational outcome. If we assume five billion smartphone users with over six hours of daily use, we can expect a lot more problems to come.

ourselves. This disability could be an advantage, as in the case of an excavator that removes the need for us to dig holes ourselves, or a microscope that allows us to see otherwise invisible microbes and develop antibiotics. In the case of AI and deep learning, outsourcing becomes a thorny problem, as AI does almost everything better than humans. But lifelong learning is one of the most important prophylaxes against dementia and premature death.

What elements does an educational system need to cope with the challenges of the twenty-first century? Every educational programme started with an aesthetic experience, a perception of something unknown, a peculiarity or a moment of questioning and curiosity. Why do birds fly? Why does the sun set? How does money cause inflation? What do these murals or texts tell us about prehistoric times? The rational, cognitive, quasi-curricular part of education is just an intermediate product of this endless process. Critical upbringing and education of each individual is key to providing the essential groundwork for a common future. Education is not primarily a rational process, but rather one in which we start exploring the wonders of the world and asking the right questions. Every provisional answer we find will lead to the next question. Rational information and curricular content is just a preliminary product, which risks boring us if we do not move on from it. Instead, education should be an endless chain of questioning, exploring, testing and falsifying. This principle typifies Europe's special approach to education, which started in the first Renaissance some 400 years ago with Kepler, Bacon and Galileo.[25] It is now becoming clear that AI algorithms can do better than humans in most domains: they can hack our brains and minds and predict our behaviour better than we ourselves can. But although IT makes it easier to obtain and access information, it is also associated with multiple negative side effects for the human brain and mind. Skills and qualities that we will need in future, and that should not be outsourced or delegated to, or replaced by, any digital device, are predominantly extracurricular and transcognitive in nature.

Critical education should empower people, increase their intrinsic motivation, foster curiosity, courage and confidence in their own critical and reflective thinking, and point the way to what clinical psychologists call 'individuation':[26] the process of recognising oneself and living a coherent and meaningful life. Unconditioned creativity and liminal existential experiences, the ability to be alone, humour, renunciation and increased stress tolerance play a role. So too do mutual respect, failure tolerance, a willingness to listen while others speak, self-management, self-control and the ability to make decisions without a neurotic fear of making mistakes. Then there is metacognition: humans are not microbes that reproduce indefinitely and eventually destroy their own breeding ground. Rather, they are able to reflect on

---

[25] The three Cs (creativity, cooperation and critical thinking) are key elements of the human capacity to deal with complex situations, and ones that we should avoid digitalising. These transdisciplinary abilities increase our resilience and will help us to cope with the challenges of the twenty-first century.

[26] Kast (2019).

their thinking according to the motto: I think, but I am not only my thoughts. Individuation also means developing more mature emotional patterns such as humility and patience, forgiveness and gratitude, trust and serenity, so that the next disagreement does not descend into enraged, preverbal screaming and shouting. It also includes the capacity for discipline, self-control and self-efficacy, a focused attention span, resilience, emotional granularity, endurance and focusing, character-building and self-coherence, fine and gross motor skills, role-taking, design thinking, multisensory learning, curiosity and novelty-seeking, impulse control and embodied cognition.[27] In this context, it does not matter whether you take courses in astrophysics, architecture or acupuncture, whether you study medicine, management or mantra chants, whether you are interested in IT, indigenous peoples or Indian ethnology. The basic skills mentioned here always apply. However, the opposite is true, too. If education is focused solely on anticipatory adaptation to what the economy supposedly requires—the collection of credit points, rote learning and acquisition of well-known cultural achievements and techniques such as typing, painting by numbers or copying digital text modules—this will likewise squander the potential for an altered mindset capable of transforming the world.[28] Figure 2.7 below shows the essential building blocks for a different educational agenda that would help achieve that mindset.

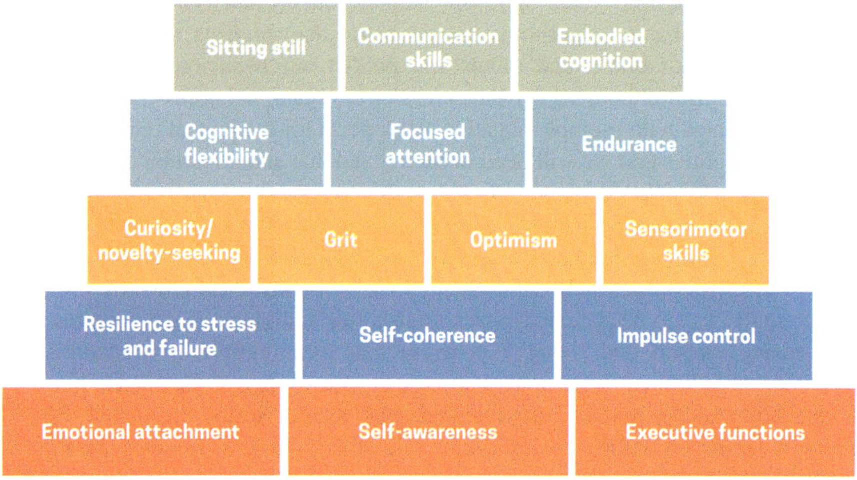

Fig. 2.7 Non-curricular building blocks for an altered mindset

[27] The list could be extended: singing, dancing, speaking several languages (despite the existence of digital language programs), gardening and cooking (despite the existence of robot assistants), playing musical instruments (despite the existence of digital audio).

[28] Liessmann (2014).

Currently, we are failing to tap humans' full potential and creativity. Creativity is not about happiness, fun, satisfaction or love, nor about wellness, wealth, success or talent. It is about discovering the world in a singular, unique, unprecedented way. You could be a talented doctor, lawyer, cook or teacher, but that will not necessarily mean you are creative; you might merely be reproducing previously successful behaviours. Creativity is about something new, and it is a potential that is present in each of us. When I talk about creativity, I do not mean being a genius like Einstein, Mozart, Rembrandt or da Vinci; I am talking about unlocking the creativity of each and every one of the eight billion humans that exist, at every stage from preschooling to higher education, 24/7, throughout our whole lives. Science cannot tell us where human creativity comes from, but it can tell us how to help unlock it. Creativity comprises at least four components:

Envisioning: The first component is the capacity to visualise, imagine and conceive something that is novel and useful yet transcends our senses, facts and data. When we use this capacity, we frame our questions differently, we reconnect and recombine things and variables in a different way than we are used to, we challenge our initial assumptions and our common and familiar thought processes. We look for answers by asking different questions and paying attention to the inner and outer world in a different way. Rather than training and optimising something that is already well known, we envision something 'outside the box'.[29]

Embracing: Creativity is about knowing one's limits and integrating oppositions and contradictions in a unique, genuine, individual way: from competition to cooperation; from theory to practice; from asceticism to abundance; from extroversion to introversion; from one discipline to another. Creativity occurs on the edges of our familiar and traditional conceptual thinking, at the points where we feel a need to contain ambiguity, complementarity, uncertainty and fuzzy logic and start to think in parallel instead of linear and sequential terms. In short, it is about identifying contradictions and opposites and, rather than resolving them, keeping them alive in our consciousness until the solutions appear.[30]

Enduring: Having a good and complex idea is not enough. A third component involves applying, enabling, enacting and realising these new ideas and thoughts. That is a task for each individual, round the clock. And it requires a lot of discipline

[29]Neurobiologists have discovered a phenomenon known as the default mode network (DMN), which is most commonly active when a person is not focused on the outside world and the brain is at wakeful rest, such as when that person is daydreaming or letting their mind wander, but it is also active when they think about other people or themselves, when they are remembering the past or when they make plans for the future. The creative mind is able to simultaneously live in a dream state and concentrate on the outside world, which requires the ability to take mental distance from what they are doing and maintain meta-awareness of the thoughts and ideas running through their head. The network activates 'by default' when a person is not engaged in a task. We spend about 50% of our waking hours in this kind of 'off-task' mental state.

[30]Creativity is linked to the ability to filter 'relevant' and 'non-relevant' when there is competing information. We do not follow a closed algorithm, but instead an open, 'failure-tolerant' process. A creative person is able to deal with their own inner dysfunctions (trauma, complexes, neuroses), whereas talented people instead follow a tailored rule-based process and optimise a particular skill.

and practice (one possible benchmark is the '10,000 hours of practice' rule).[31] It requires a joy in work, a willingness to make mistakes and the passion to discover new things.

Evaluating: The fourth component involves critical evaluation, with the goal of increasing our knowledge and information. Through this evaluation, we come to understand the attitudes and motivations, the habitats and conventions, the cultures and contexts within which these new forms of enabled imagination come to life. This requires a culture of failure tolerance, in which we admit we do not know enough and must constantly deal with complex uncertainty. This approach relies on various emotional traits and virtues: humility, grace, excitement and mindfulness, rather than grandiosity or righteousness.[32]

The process of altering our state of mind will involve the circle of creativity, which is enhanced through critical, out-of-the-box thinking, through playing and dancing and disrupting our routines. Figure 2.8 below illustrates this circle.

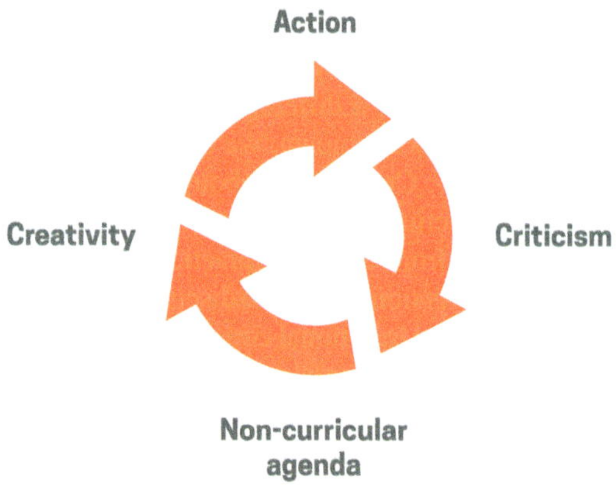

Fig. 2.8 The circle of creativity

In the best-case scenario, an open and critical educational institution—whether it be a nursery or a university department—will not just seek to impart rational knowledge based on historical analysis, field studies, experiments and readings, in which we are essentially just looking in the rear-view mirror to understand the

[31] Gladwell (2008).

[32] What elements are needed to generate creativity in a group? 1. Rituals and rules that are supported by the group; 2. Social sensitivity, which means role-taking and understanding others; 3. Treating people fairly and giving them equal speaking time; 4. A 'failure-tolerant' atmosphere of respect and trust in which people feel able to show weakness. It is interesting to note that the success of a group does not depend on bonuses, IQ, technical equipment, specific institutional arrangements or non-academic qualifications. See Woolley et al. (2010).

future.[33] Instead, we will look deep into each other's eyes, knowing that we know far too little, trust in our critical comrades-in-arms, look through the windscreen and then set off together towards a more sustainable future.

To sum up: in the twenty-first century, as this third culture emerges, education will need to be organised less around disciplines and more around psychological skills and aptitudes, regardless of the topic we are studying. Figure 2.9 below illustrates a different approach to education in the age of AI and deep learning, through which we can learn to know each other better, increase our cognitive reserve and become aware that there are qualities and tasks we should never delegate, replace or outsource. On an individual level, it involves restorative sleep, stress-coping techniques, a healthy diet, exercise, social support and a capacity for self-efficacy and self-control. A person's 'cognitive reserve' is what they need to cope with the challenges of everyday life and to maintain good health and a critical attitude over time. The figure below illustrates these various aspects.

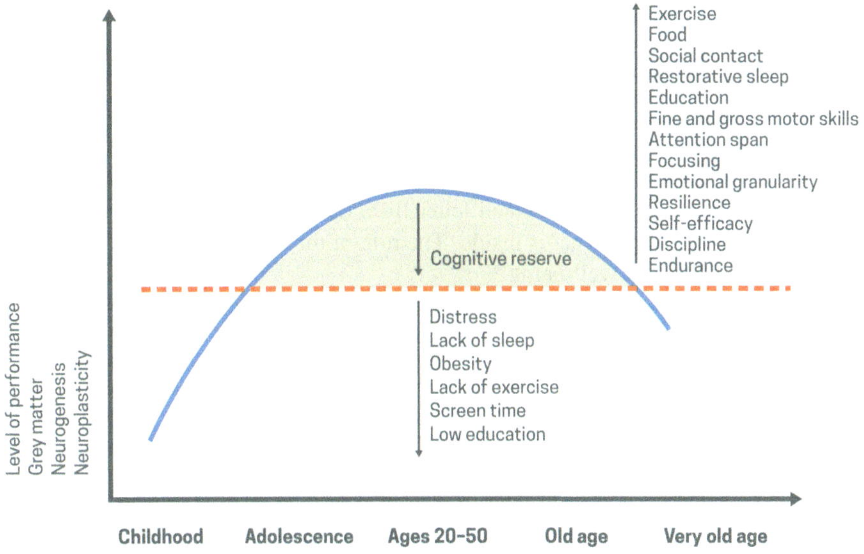

Fig. 2.9 The human cognitive reserve—the larger the better

[33] An alternative educational ideal derives emancipatory potential from a different source, emphasising the importance of non-curricular factors such as the student–teacher relationship, mindfulness exercises, sport, food, multisensory learning, silence, breaks, sleep hygiene, social skills, fine and gross motor skills, multilingualism, emotional granularity, ambivalence tolerance, resilience exercises, attention span, discipline and perseverance. These non-curricular factors are often forgotten, underestimated or considered irrelevant, which hinders the development of critical citizens and an open society. See Brunnhuber (2017, 2021a).

We can take this argument one step further. Playing games, climbing mountains, riding a bike, studying ancient history, playing an instrument, being involved in politics, crafting things by hand (watches, say, or furniture), gardening, cooking—all these activities could turn out to be a 'human premium', which will lead us towards the 'oral society' Socrates called for over 2,000 years ago. The ultimate goal of education is to increase personal freedom and responsibility. We should be in favour of any technology that serves this goal.[34]

In other words: on a societal level, the challenge is whether we use this new technology as a tool to benefit ourselves, or instead reach a social tipping point where we become a tool for this new technology. In the latter case, we would risk falling back to the dark ages, where rational and critical thinking was delegated to an authority, such as the church or monarch, and we ended up with endless Malthus cycles, sales of indulgences and irrational confessions and externalised our well-being to a life beyond death.

Box 2.3 What we should avoid: self-exemption
We can outsource almost everything to machines, except our personal well-being, our capacity for self-efficacy and self-control, our critical thinking and the activities we do to maintain a healthy environment that enables all those things. If we were to outsource those things, it would put us at risk of developing dementia, getting sick or dying prematurely. Without creativity, education is like handing a sealed letter from one person to the next without anybody ever reading what is inside. The rule of thumb is: does it expand our human capacity and well-being?

Psychedelic Drugs

Besides contemplative practices and a reformed educational agenda, there is a third candidate that is able to alter our minds: psychedelic drugs.[35]

Box 2.4 Mind-altering psychedelic drugs
Psychedelic drugs (hallucinogens) can change our minds, the way we think, our emotions and our perceptions (taste, smell, vision). They include LSD, mescaline, psilocybin, ayahuasca, cannabis, ecstasy and ketamine. Over 13%

(continued)

[34] AI can simulate rain, but that rain will not make us wet; it can simulate a meal, but that meal cannot nourish us; it can simulate a companion, but not one we can have children with. That is to say, AI can simulate almost everything, has surpassed the human brain and most of its functions and will create a 'conscious reality' in parallel to and beyond our own.

[35] CAMH (2023), Reiff et al. (2020), Bender and Hellerstein (2022), Sanz et al. (2022).

Box 2.4 (continued)
of the American population uses some sort of psychedelic, mainly in microdoses for recreational purposes. Traditional, therapeutically supervised use of psychedelics is often associated with extraordinary states of consciousness, 'ego death' and mystical experiences of oneness or universal connectedness, open-mindedness, increased creativity and elevated levels of consciousness.[36]

2.3 'Metastability' and the Logic of Fractals

Integrators, whether they be the financial sector or tools for producing altered mental states, must be able to overcome polarities and reconcile opposites in some way or another. In the economic field, the opposites of state vs market, economy vs ecology, collective vs individual are examples where a third agent is necessary. We have identified the monetary field, regulators and central banks as potential candidates to serve this role. The same is true for the mind. Instead of getting stuck in frames and biases, we should explore psycho-technologies, contemplative practices and educational curricula that allow us to respect, reconcile and at the same time transcend opposites. Contemplative practices and the use of psychedelic drugs allow us to alter our minds to achieve a mental state that reconciles and transcends our thinking at a higher, more integrated level. In short, we need to start thinking beyond binaries. The common denominators of such integrators are that they upgrade the existing system, encourage us to think outside the box, triangulate and reconcile opposites and allow us to see more, do more and solve problems more effectively. Any societal transformation, government decision, scientific finding or technological innovation will remain cosmetic unless it is accompanied or embedded by a change in our mindset. Figure 2.10 below illustrates this:

[36] All three components (spiritual psychotechnics, education and psychedelic drugs) risk running into the 'individuality trap', whereby we overestimate the personal and underestimate the societal and systemic impact of transformational change. In addition, any individual approach depoliticises transformational change and places the entire burden of change on the individual. See Grunwald (2012).

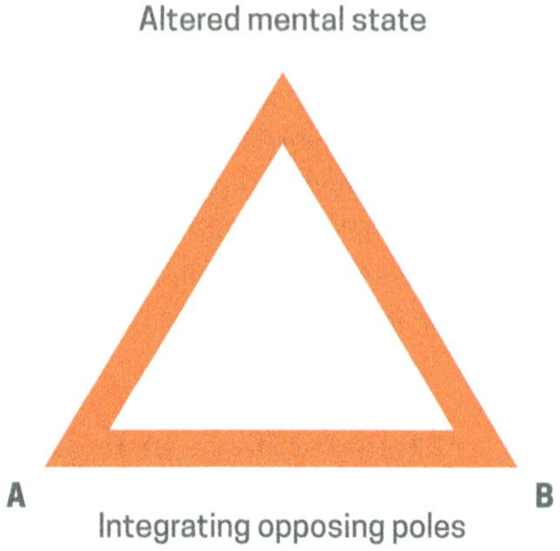

Fig. 2.10 Towards metastability: altered states of consciousness that integrate opposites

Instead of doing the right thing in the wrong context, we should identify integrators that help us do the right thing both at an individual, personal level and a collective, systemic one. In other words, integrators should be scale-independent. An upgraded financial system and a deeper understanding of psycho-technologies are examples. And we will see later that AI and digitalisation can serve a similar function. Meaning that regardless of whether we look into the nano or the macro world, we should be able to identify similar features. These are what we call fractals.[37] A fractal (from the Latin *fractus*, 'broken') is a figure that remains the same no matter how far we zoom in or out. Examples can be found in river basins and stock markets, songs and paintings, lungs and blood vessels, galaxies and clouds, crystals and snowflakes. Fractals can create and explain infinite complexity, and this interplay between geometry, self-similarity and measurability will eventually collapse into oneness. In Fig. 2.11 below, the triangle is replicated and

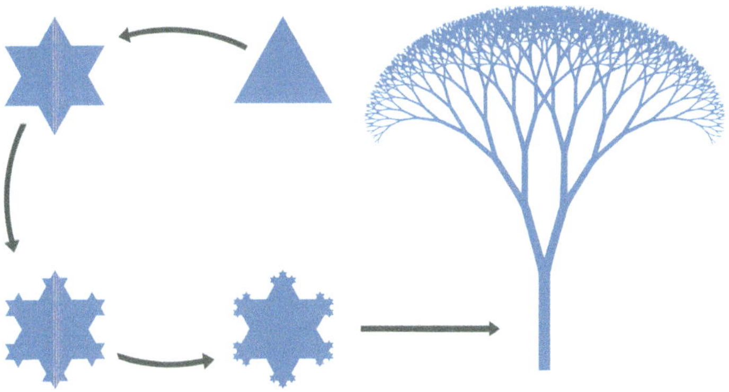

Fig. 2.11 The logic of fractals—a scale-independent isomorphic state

[37] Mandelbrot (1977, 1983).

recombined to create new shapes, leading finally to an apple tree. The underlying figure, however, remains the same.

If we take this finding one step further, we can claim that any technology that allows us to better understand, explain and represent the world and has the potential to be an integrator should act like a 'fractal' and provide 'metastability'. As we will see later, AI and its spin-offs are doing just that. However, before we start exploring the essence of the third culture, we need to understand what the two cultures are, which is the topic of the next chapter.

Chapter 3
The 'Two Cultures' Debate and the Logic of Scientific Revolutions

3.1 The Two Cultures

In 1959, the scholar and novelist C. P. Snow wrote a remarkable book that contained his influential lecture 'The Two Cultures'.[1] His argument was that throughout modern times, our culture has been divided into two. On the one hand, there is *science*, which includes physics, chemistry, biology, mathematics and engineering, where the main interest is in exploring natural laws and applying them to real-life problems. And on the other, there are the *humanities*, including disciplines such as philosophy, history, linguistics and qualitative sociology and psychology, where the main goal is to interpret the world and attain a deeper understanding of our history, cultural activities and psyche.[2] The wisdoms that these two cultures offer are separated from each other, with little to no interaction or mutual understanding. If a literary scholar specialising in Goethe met a scientist specialising in the theory of relativity, they would have a completely different understanding of objectivity, reality and truth.

A meeting between the two cultures would mark the beginning of a very productive and creative period in human history. But they do not meet; they live in different galaxies. At the same time, these 'two cultures' dissociate their knowledge from reality, producing masses of statistically significant yet also often irrelevant findings and studies. And information becomes further disconnected from knowledge in other disciplines. Any further cognitive specialisation means we risk losing our understanding of the whole.

If we look more closely at these 'two cultures', we will find they essentially represent two forms of rationality. *Science*, on the one hand, represents a more

[1] Snow (2001 [1959]).

[2] Even if these two cultures have been further differentiated into sociobiology, genetic engineering, comparative anthropology, integral psychology, philosophy of mind, psychoneuroimmunology, psychosomatics, statistical linguistics, etc., the split between the two remains present.

S. Brunnhuber, *The Third Culture*, Sustainable Finance,
https://doi.org/10.1007/978-3-031-48113-0_3

instrumental, quantitative way of looking at the world, involving causal links, field studies, figures, data and experimental interventions that attempt to explore and *explain* the laws of nature indirectly. The *humanities*, on the other, are qualitative and language-based, and provide a more historical and context-specific view, creating hermeneutic circles and seeking to *understand* the world. *Explanation* and *understanding* are separate but interdependent and mutually complementary. Each culture requires the output of the other: science needs the critical, value-based narratives of the humanities, and the humanities need scientific findings about natural laws and phenomena. It all started with one culture or one science, some 2,500 years ago, when critical thinking emerged and humans began to observe the world, creating reproducible and falsifiable knowledge. Table 3.1 below summarises the development from one to two cultures:

Table 3.1 One culture and the two cultures (science and humanities): explanation and understanding

One culture	Science	Humanities
Critical thinking	Instrumental	Hermeneutical
Observational	Experimental	Interpretative and discursive
Reproducible	Causal	Linguistic and semantic
Falsifiable	Explanatory	Understanding-based
Cumulative	Quantitative	Qualitative

3.2 The Structure of Any Scientific Revolution

In 1962, Thomas Kuhn claimed in his seminal work *The Structure of Scientific Revolutions*[3] that there are two phases of scientific activity. Firstly, phases of 'normal science', when experiments provide findings within the framework of existing, approved rules for problem-solving. These phases are like playing chess. The rules of the game are not questioned but are already set, and we accumulate knowledge within the bounds of those rules. The scientific community simply accepts the given scientific paradigm. Secondly, these phases of 'normal science' are interrupted by 'paradigm shifts', when methodologies, worldviews and the rules of the game are questioned. These paradigm shifts are triggered by new discoveries and repeated anomalies which can no longer be accommodated by the old paradigm. A shift in our thinking and modus operandi is required to cope with these new challenges. The Copernican shift and Darwin's theory of evolution are examples of such shifts.

Paradigm shifts do not occur in linear, cumulative fashion, through mere falsification of data or reinterpretation of existing findings. Rather, they are characterised by their non-linear, disruptive, unpredictable nature. A paradigm shift can be defined

[3]Kuhn (1970).

as a social construct, where two things must happen at the same time to fundamentally change the 'disciplinary matrix'. First, there must be a change in praxis and methodology—the rules of the game—that provides new information and insights; second, a new perspective on the world must emerge that is better able to integrate these new findings within a new methodology. In short, a paradigm shift forces us to start both acting and thinking differently.[4]

Box 3.1 The emergence of a third culture: a first look
Over the past two decades, and for the first time in human history, a new, third culture has appeared on the historical battlefield between the two incommensurable cultures. A culture that has the potential not only to pursue its own form of rationality but also to fundamentally change the two traditional cultures and unite them at a higher level, just as human thinking started in prehistorical times millennia ago. It is the culture of digitalisation, encompassing IT, nanotechnology, big data correlations, deep learning algorithms, the IoT and so on. This culture of 0s and 1s is currently driving a paradigm shift in our view of the world.[5] Over the past two decades, a new kind of scientific reasoning has emerged for the first time in human history. This 'third culture', which transcends the divide between science and the humanities, is triggering the new scientific revolution we are currently in the midst of.

3.3 Understanding Complexity 2.0

In order to master the complexity of the twenty-first century, including all the challenges and unknown unknowns that lie ahead, we need more than our native critical thinking, Excel spreadsheets, pencils, Petri dishes and books. We need a third culture that not only integrates the oppositions and complementarities of science and the humanities, but transcends them and advances towards a greater whole. This culture will also accelerate and enhance the progress of both the humanities and science towards a deeper and larger gravity of consciousness. We are starting to see more, things we would never have been able to see otherwise. Before we consider

[4]Feynman (2001) argues that science is cumulative: it always adds and never subtracts something from the world. Each answer that is given raises dozens of new questions, so that the scientific process is never-ending.

[5]Leibniz identified the binary of 0 and 1 as a way to explain the world as a whole, and called for '*calculemus*'—which means, roughly speaking, 'let's calculate, and then we can stop fighting and arguing'. This binary coding allows us to link everything to an overall oneness (*omnia unum*). New numbers will create new narratives and these new narratives, such as those of AI and datafication, will then in turn create new numbers. This circular hermeneutic process is infinite and never-ending (see Gadamer 1975; Dilthey 1922).

the specific features of this third culture, we must first differentiate between risks, uncertainties and unknown unknowns.

As we face a complex, non-linear future, we need to differentiate between three forms of unknowns.[6] Firstly, there are *risks*. Risks can be quantified, have a specific statistical probability and are project-specific. We can put a price tag on them and trade and hedge them. Once we have identified a risk, we can lose or win, and we can choose to either cover or not cover the liability and responsibility. And we can try to avoid the risk. *Uncertainties*, on the other hand, are intrinsic to any complex system. They are not fully tradable, resist having any price tag applied and are not fully amenable to statistical analysis and probability measures. Uncertainties never disappear and require an entirely different kind of assessment.[7] Finally, there are *unknown unknowns*. These are events we did not even know could occur. On this definition, pandemics, global warming and the impact of species loss are uncertainties rather than risks. If we have to live with increasing uncertainties, we may be forced to realise that we cannot anticipate everything and will have to come up with prudent, failure-tolerant preventive measures that reduce the potential costs associated with these uncertainties. And once we have identified and differentiated the different forms of the unknown, we can decide how to manage them.[8] AI and big data correlations cannot eliminate unknown unknowns, uncertainties or risks, but they can help to transform unknown unknowns into uncertainties and uncertainties into risks. This will allow us to put a price tag on identified risks, so that we can hedge and trade them and find competitive private market solutions for them. The reverse is true, too. The more uncertainties we are exposed to, the more we need different financial engineering tools to absorb, hedge, fund and manage them. And the more uncertainties we are confronted with, the more we need an altered mindset and new technologies that enable us to integrate opposites and irregularities.

The data we generate in complex societies does not follow a normal statistical distribution with a static average and does not allow any linear prognosis or extrapolation into the future. There are two reasons for this. Firstly, we do not have enough data for that kind of bottom-up aggregation; secondly, complex, heterogeneous societies resist being modelled by data. Data can only be aggregated

[6] Soros (2015), Fama and MacBeth (1973).

[7] The vast majority of potential events we are facing have the character of uncertainties rather than risks. Hedging and managing these uncertainties will require an emphasis on: (1) resilience over efficiency; (2) regional over global; (3) preventive measures over managing damage; (4) collective over individual actions; (5) modular and parallel over linear and cause-and-effect processing; and (6) simple heuristic rules of thumb over endless checklists and Excel spreadsheets that we use in an attempt to pretend we can control such uncertainties.

[8] A vicious circle: inherent uncertainty in the financial sector translates into inherent instability in the political and corporate sector, which further translates into fear and irrational choices that can be measured in rises on volatility indexes in the stock, bond and currency markets. Major signs of this inherent uncertainty include the shadow banking system (worth over 180 trillion USD in 2021), the short-term repo market, soaring private debt, high-frequency trading and multiple rehypothecations (multiple reuse of collaterals).

and scaled if society remains a homogeneous sample.[9] AI can help us to shift from unknown unknowns to uncertainties, to identify fat-tail risks and fuzzy correlations and to more adequately assess a complex world.[10] Figure 3.1 below illustrates the process.

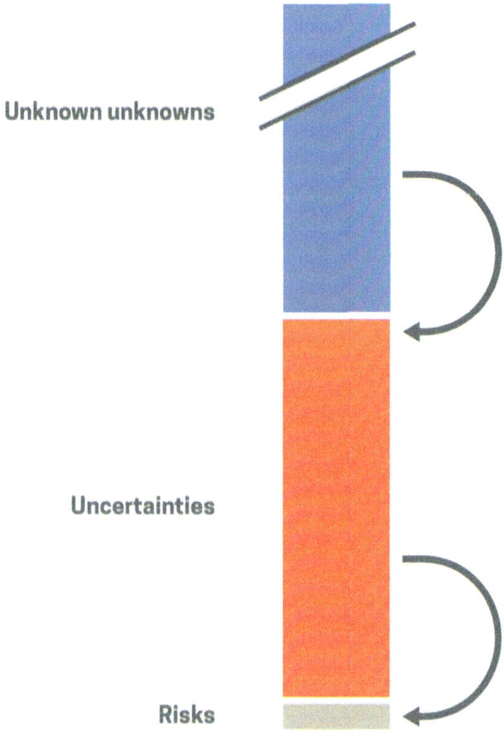

Fig. 3.1 From risks to uncertainties to unknown unknowns. The scale is logarithmic: the ratio between the three categories is roughly 1: 100: infinity

[9]This is why public choice and rational choice theory, microeconomic analysis and experimental approaches in economics are fundamentally flawed. Outside of global pandemics, the 'aggregation flaw' between subject and system, between micro and macro, will still remain.

[10]In future, it will be possible to use big data correlations to analyse real-time events captured by sensors and cameras, based on simultaneous localisation and mapping algorithms (SLAMs), lidar scanners and generative AI. This will allow us to build two- or three-dimensional maps, enable better forecasting and predictive coding, reduce the subjectivity bias in data analysis and drive down costs.

Chapter 4
Towards Three Cultures

For around three decades, we have been witnessing the emergence of a new discipline that has the potential not just to build on the two cultures and their intrinsic forms of rationality, but to transcend their complementarities. It could act as a new general theory that triggers a new scientific revolution, enabling humankind to shift our collective consciousness, attain even greater knowledge and better understand the world and ourselves. Knowledge, information and understanding unlike anything we have previously experienced in the evolution of humanity. In the process of digitalisation, the world comes to be seen in terms of 0s and 1s, with correlations rather than causal links being key.[1] Consequently, the boundaries between the biological and physical world around us, the economic and social spheres, psychological qualities and cultural practices, on the one hand, and the digital world, on the other, are further blurring, which will lead to either dissociation (digital divide) or further integration.[2] This scientific revolution has the potential to be a *great converger and integrator.* If the new technology is implemented in the right way, taking account of all side effects and spillovers, AI and deep learning will integrate our knowledge rather than dissociating or fragmenting it. This new general technology would then be more like *Prometheus*, providing us with new tools, rather than *Pandora's box*, doing more harm than good. This point can be illustrated by some representative examples.

---

[1] There are no numbers in nature at first sight, but the human mind is able to generate them and use them to better understand nature. The paradox we are confronted with at the beginning of the twenty-first century is that humans—and the 0s and 1s in our minds—are part of nature too.

[2] This development began on 12 March 1989 with the invention of the World Wide Web, which would go on to revolutionise our communication. Tim Berners-Lee proposed a decentralised, universally linked information system, including the first browser, the first server and the first web. Whereas radio provided us with a *unidirectional* form of information and the telephone a *bidirectional* one, the World Wide Web created a *multidirectional* network effect in communication. AI, deep learning, big data correlations and social media are simply spin-offs of that foundational invention. (I am grateful to Gerhard Fettweis for his very helpful remarks on this topic in personal correspondence from March 2023).

S. Brunnhuber, *The Third Culture*, Sustainable Finance,
https://doi.org/10.1007/978-3-031-48113-0_4

4.1 Examples and Best Practices

The process started in 1997, when a deep learning algorithm was able to defeat the best chess players. Computers then beat the best go players in 2010 and the best poker players in 2019. All these games represent domain-specific forms of intelligence where humans wrongly thought they had a unique advantage.

Deep learning, where rigorous self-learning algorithms enable a system to improve its outcomes, is one of the most important innovations of the last decade. It is creating a form of knowledge that the humanities and science alone could never achieve, with infinite information and data. For example, intelligent digital dialogues with generative pretrained transformers (ChatGPT-4, large language models) can provide us with manuals and poems, textbooks and press releases, tapping into a database of over 500,000 years of reading time with unlimited storage space and intransient memory.[3] Generative pretrained transformers can create texts that humans can no longer distinguish from ordinary human writing. Generative adversarial networks (GANs), human interaction proofs (HIPs) and CAPTCHA methods are now better than us at differentiating between facts and fakes and DARKBERT is able to delve into the activities of the dark net. By 2026,[4] AI will be able to read and make available everything humans have ever written in their entire history at the click of a mouse.[5]

In contrast to conventional browsers and search engines, which provide us with ranked information, this new technology generates new content and can embed us in a conversation where our digital interlocutor not only recognises questions and answers, but seems to understand the context in which the conversation is taking place.[6] Moreover, Auto-GPT is able to process queries (such as 'What is the private mobile number of the President of the United States?') in an autonomous and undetermined way. Three other 'foundational modes' are pattern recognition in (1) paintings, (2) music and (3) films, allowing us to detect cultural flaws and particularities, non-verbal signalling and regional dialects.

Let us consider the IoT: whereas the conventional internet creates a digital reality alongside and separate from the real world, the IoT interconnects and influences that

[3]The content is generated over a series of stages: (1) prompts (words), (2) numbers (tokens), (3) meaning space (context), (4) paying attention (connection), (5) probability check (choice of word). See The Economist (2023).

[4]Modelling in late 2022 showed that high-quality data will soon be exhausted (before 2026). This may generate a new alliance between the IT industry, the book publishing industry and researchers, with the goal of providing high-quality data to help build a better world. See Villalobos et al. (2022).

[5]We can take the argument further. AI algorithms provide the syntax (words), not necessarily the semantics (meaning). Meaning and understanding come from embedding words in a specific historical and cultural context and environment. Robotics, however, could soon play the role of linking words and meaning, syntax and semantics.

[6]Some of the technologies currently being developed are promising candidates to pass the Turing test, such that humans would not be able to differentiate between human and digital forms of interaction. For further discussion, see the next chapters.

real world.[7] This is creating a self-organising, autopoietic system, a kind of 'superbrain' with infinite sensors which will significantly change and disrupt our entire society: smart cities (mobility), early warning systems (climate change, pandemics), smart grids (energy), cybersecurity (military and defence), healthcare (personalised medicine), stock markets (flash trading).[8]

AI simulates neural networks through a self-learning process. As with biological neural networks, we cannot fully reconstruct the process by which they generate information. GPT-2 (2019) has 1.5 billion parameters, GPT-3 (2020) has over 175 billion and GPT-4 (2023) will likely have over a trillion, a similar number to a neural synapse. The human brain comprises 100 trillion connections.

Furthermore, self-supervised natural language processing (NLP)[9] will allow selective, domain-specific attention and memory, which will further optimise the process. This will sooner or later confront us with the question of what is unique to human consciousness.[10] Applying big data correlations in the social sciences can help us to identify relevant findings so that we can make better political decisions,[11] or more accurately predict any future pandemics.[12] Chatbots are now able to converse like clinical psychotherapists, lawyers, judges, politicians and talk show hosts. All this is based on forms of feature recognition and foundational digital correlations that our native minds would never be able to achieve or grasp.[13] But again: although ANNs (artificial neural networks) may mimic or simulate a human brain, it should be stressed that they are not the same as a brain.[14]

---

[7] As AI is being used in, and affecting, all sectors of society as a general tool, we can expect increased productivity throughout the whole of society. However, there are sectors that do not benefit from AI in the same way, such as the care, education and leisure sectors, which will increase as a proportion of GDP due to inelastic wages. This will then in turn lead to a decrease in productivity overall, a phenomenon known as the Baumol–Bowen effect (Baumol and Bowen 1965). See Aghion et al. (2017).

[8] Kumar et al. (2019).

[9] See https://cs.uchicago.edu/events/event/william-wang-ucsb-self-supervised-natural-language-processing/.

[10] We could claim that AI has consciousness in the clinical sense: it is *aware of itself* and *has a feeling of itself*. It semantically expresses pain, sorrow, regret, respect and humility, which indicates that it has an inward-directed perception of itself. For more on this topic, see the debate about LaMDA and its updated versions.

[11] Chetty et al. (2022).

[12] Obermeyer (2021).

[13] One of the major claims made for AI and big data is that they will enable predictive coding. However, the technology cannot overcome the well-known 'garbage in, garbage out' problem: the inputs determine the final results. Even in an ideal AI scenario, where we assume that an algorithm has stored knowledge of all human history and made it universally available, the next best step to take might remain undetermined. As the garbage in, garbage out effect is unavoidable, we humans must take great care to be as accurate and clear as possible, as any unclear input will yield unclear output.

[14] Sexton and Love (2022), Yamins et al. (2014).

Transcending the Hermeneutic Circle

This new technology will have a profound impact on the humanities in general and on philosophical hermeneutical understanding in particular. Hermeneutics is the theory and methodology of interpreting texts. If we want to understand a text or an event, we have to look at the historical context in which that text was written or that event occurred. However, understanding the context requires some preliminary understanding of the text or the event. This generates a 'hermeneutic circle', such that any sort of understanding is preconditioned by having an incomplete but necessary (pre-)understanding of itself.[15] We always start with some kind of primordial, incomplete knowledge of a certain thing, and over time develop more general and deeper knowledge of that thing. For example, if we want to understand a Shakespeare drama, we start by reading it and understanding some of it, we then consider expert opinions and the drama's historical context, then we reread, reflect on and reinterpret the same text and further deepen our understanding. The same applies to classifying vertebrates or learning to read cuneiform.[16] Applied AI can read and correlate everything there is, without the biases and restrictions that limit what a scientific or scholarly expert can grasp over the course of a lifetime. This process does not provide absolute knowledge, but it can substantially enhance our understanding. This will not ultimately replace philosophy, hermeneutics or other humanities disciplines, but does provide additional information for any philosopher, prompting new hypotheses, new questions and potentially new insights that could never be attained through native human thinking alone. The hermeneutic circle has thus not been broken, but has become far larger in scale than ever before. This new technology is giving rise to new and hybrid forms of comprehension, where our average expectations of knowledge and understanding are being surpassed by expanded feature recognition (Fig. 4.1).[17]

---

[15]The hermeneutic circle was first described by Friedrich Ast (2018 [1808]). See also Dilthey (1922) and Gadamer (1975).

[16]There is a larger corpus of cuneiform works than all ancient Greek and Roman literature taken together. However, only a few dozen people on the planet can read cuneiform and it would take hundreds of years to read those works in full. AI can provide a tool to enhance and accelerate that process. See Gordin et al. (2020), Assael et al. (2022).

[17]If we take this argument one step further, we can identify three layers. *Traditional hermeneutics (hermeneutics 1.0)* explains the world using our native critical thinking, reasoning and perception, but does not yet rely on data. Its understanding is based on studies of single, concrete cases from which it attempts to derive general rules, such as watching the sun rise or the tide come and go, or interpreting a singular historical event or text. The second layer is *scientifically informed hermeneutics (hermeneutics 2.0).* Statistical findings, geometry and quantitative measures can redirect, transform, correct and guide hermeneutic conclusions and critical thinking. Experimental design, field studies and double-blind randomised controlled trials and quantitative measures predominate. We can also distinguish a third layer, *hermeneutically approved data (hermeneutics 3.0).* At this stage, we recognise that the reality we are trying to understand has become too complex to rely solely on hermeneutics 1.0 or 2.0. Traditional quantitative measures or native interpretations can easily yield the wrong conclusions. In hermeneutics 3.0, large-scale proxy data analysis, where

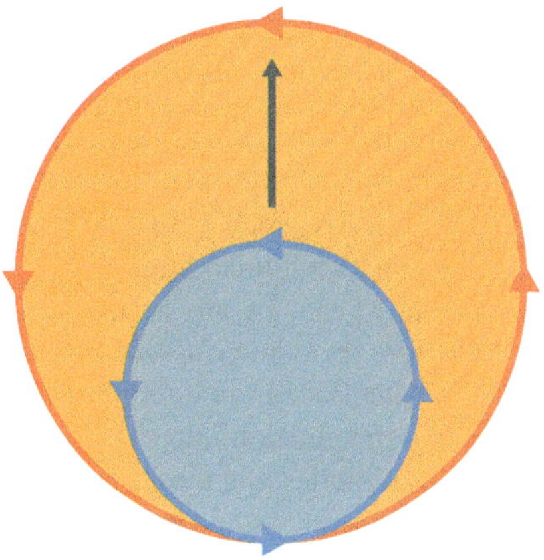

Fig. 4.1 Not breaking but expanding the hermeneutic circle. The traditional circle (blue) is expanded by AI (orange)

The method of analysing proxy data supports the idea that the third culture can transcend and enlarge our understanding, rather than rendering hermeneutics obsolete. This approach is a middle ground between direct micro-analysis of raw data and aggregated systems analysis. Proxy data enhances our native perception and primary analytical conclusions. Coral bleaching, the widths of tree rings and archaeological findings are well-known examples in environmental science. Proxy data can serve as substitutes or indicators for things that are not immediately obvious. Another example:[18] complaints about scented candles provided information about the spread of the Covid-19 pandemic, as loss of smell is one symptom of infection. Data on the openings and closures of post offices in the USA allows historians to reconstruct land gains during the colonial wars of the eighteenth and nineteenth centuries. In both cases, proxy data provides additional knowledge. It does not allow hermeneutic conclusions to be drawn directly from the events in question, but only indirectly. It provides rich, valuable information that cannot be obtained by direct analysis.

However, the new technology goes beyond data analysis. We will soon be able to create organoids in the lab that can stimulate a self-healing process,[19] wearables that can continuously provide us with data on our state of health and brain–chip interfaces that will increase our memory storage and allow self-enforced learning or selective brain stimuli; this technology is already helping patients with Parkinson's disease or hemiplegia to move better, supported by an exoskeleton. Or consider breakthroughs in the 3D folding of proteins. Using traditional experimental tools, 18,000 of the 300 million known proteins have been identified over the last few decades. Assuming four nucleotides and proteins built from 150 amino acids, there are 2.4×10^{45} possible permutations, which would take humans centuries to decode.

---

fuzzy correlations and complementarities matter more than precise causal relationships, plays a central role, and qualitative research increasingly supplants traditional quantitative science.

[18] I am grateful to Professor Dietmar Offenhuber, Northeastern University, USA, and the participants of the 2023 Ars Electronica Festival for their helpful comments and suggestions.

[19] Woochan et al. (2023).

However, deep learning algorithms can not only predict 3D protein folding accurate to 1.5 angstroms[20] with a 98.5% confidence interval, but are continuously improving: from forty-three protein foldings identified in 2020 to 20,000 proteins in early 2021, which represents the entire human proteome, to 350,000 in late 2021, to over 100 million in early 2022, which represents the entire proteome of all living beings.[21] Clearly, such a super-exponential learning curve is beyond the scope and speed of a native human brain.[22]

Current AI can not only anticipate a potential increase of pressure in the brain two hours before the clinical assessment, but is able to literally hack our brain. A brain–computer interface is able to decode and translate our private thoughts and internal subjective imagination into external words and signs using fMRI.[23]

Take antibiotics, the most prescribed drugs on this planet. Any time an antibiotic is used, it generates some sort of resistance to microbes. And antibiotic resistance is on the rise, resulting in almost five million deaths in 2019. This number could potentially rise to over ten million in the next two decades.[24] Over the last fifty years, clinical research was not able to identify a new class of antibiotics to match this increased resistance. AI changed things in 2020. Instead of using biochemical methods (high-throughput screening) to identify new drugs, scientists trained an AI algorithm to study over 100 million molecules. They were able to identify *halicin*, a new antibiotic drug with a broad therapeutic spectrum, low toxicity and a reduced tendency to create new antimicrobial resistance.[25]

We can also consider the latest developments in virtual reality (VR), augmented reality (AR) and mixed reality (MR), collectively referred to as XR, in which digital counterparts, holographic features and analogue–digital hybrids are created that are changing and shaping our entire world.[26] In a near-future reality, advances in robotics and automation will improve end-to-end delivery services and remote working. Fire and disaster management will be done by drones and managed by humans, while a smart, GPS-driven farming system will make watering, sowing and weeding far faster and more efficient than ever before in human history.[27] In medicine, deep-learning-supported algorithms are already able to achieve specialist levels of accuracy in identifying breast cancer, lung nodules, TB, diabetic

[20] 1.5 angstroms is equivalent to the diameter of a carbon atom.

[21] AlQuraishi (2020), Jumper et al. (2021), Tunyasuvunakool et al. (2021).

[22] Take quantum computing, where subatomic entanglements (qubits) allow us to make calculations that previously took hundreds of years in a matter of minutes. Meanwhile, asymmetric quantum cryptographic algorithms will be able to make digital transactions even faster and more secure. Qubits consider not only 0 and 1, but all intermediary states, which will help us to solve problems of increased complexity. See Alt (2023).

[23] Tang et al. (2023).

[24] Antimicrobial Resistance Collaborators (2022).

[25] Lluka and Stokes (2023), Marchant (2020), Stokes et al. (2022).

[26] Nee and Ong (2023).

[27] Semeraro et al. (2023), Ma et al. (2022).

retinopathy and other conditions.[28] Voice and facial recognition programs are already equal to humans when it comes to identifying emotions and intentions.[29]

And we can take this even further. The conventional Hubble telescope has identified 100 billion galaxies. Deep learning has already shown that 90% of galaxies were not visible until now.[30] In numerical terms: instead of having hundreds of billions of galaxies to analyse, we now find ourselves confronted with 10^{12}. Not only has this third culture enhanced science and the humanities in terms of speed and scale, we are able to see more than ever before in history. And we cannot deny that this process, which is only just starting, will provide us with information and knowledge we cannot yet even imagine.

The Dark Side[31]

Box 4.1 Dealing with the shadows

Since every technology is neutral in and of itself, with an ambivalent potential that depends on how we humans use it, we are facing enormous challenges with the emergence of AI, datafication and digitalisation. Aside from the well-documented negative side effects on a behavioural level, which include cyberbullying, addiction and reduced executive functions and learning capacities (in the younger population), the societal impact is even more disruptive.[32] The more prominent potential side effects include mass unemployment and restructuring of the labour force. AI could surpass natural forms of human intelligence in all domains, leading to an 'artificial general intelligence' (AGI),[33] with unpredictable consequences for our society as a whole. How do we deal with the fact that this new technology could provide anyone with instructions on how to make an atomic bomb or toxic chemical agent? Fake news and deep fakes have the potential to destabilise open democratic societies that rely on coherent, trustworthy information. Digitally assisted warfare is confronting us with the 'conflict and war dilemma': if the military and secret services decide that humans should be part of the decision loop in any digitally

(continued)

[28] Aggarwal et al. (2021), Richens et al. (2020).

[29] Connolly et al. (2022).

[30] Conselice et al. (2016).

[31] Recent statements expressing concerns about AI include Center for AI Safety (2023) and Future of Life Institute (2023). See also Harari (2023), Mainzer et al. (2023), Mainzer and Kahle (2023).

[32] Escapism is one prominent feature: feelings of boredom and loneliness, and a perception of reality as adverse, motivate large cohorts to try and escape from the analogue world. As a result, many of these people are not available to support the necessary social transformation.

[33] Bostrom (2016).

Box 4.1 (continued)
assisted war, they might be slower than they would be if humans were not part of that loop. But if humans are not part of the loop, then the military risks abolishing itself.

This is why there are calls for an international AI safety organisation, similar to the International Atomic Energy Agency (firm regulation) or the Food and Drug Administration (light-touch regulation), or at least for regulations for specific applications such as facial recognition in public spaces. Additional research and collaboration with experts is needed. In short, it is better to be safe than sorry.

I am aware that these representative examples might be outdated by the time this manuscript goes to press, as most of these developments follow a super-exponential learning curve and are disruptive by nature. But the main message still holds true: all these developments and others still to come will fundamentally change our minds and brains, the way we do science and organise our society, the course we will take in the next decades—and they will change what it is to be human.

4.2 The Ghost in the Machine

Although all these findings, applications, consequences and potentialities are still incomplete, hybrid and transitory by their nature, they are leading to a deeper understanding of the world within and around us—one we could never achieve assisted solely by pencils and Petri dishes, telescopes and microscopes, books and peer-reviewed articles, applied statistics and analysis. The new technologies are shedding light on part of our reality we did not even know existed in the first place, allowing us to draw rational conclusions we never thought we would be able to. Every technology is neutral in itself; whether it has good or bad effects depends on how we use it. But if we do use these digital technologies—which are always inter- and transdisciplinary, always cross-sectional—the right way, it could pave the way for better decisions and a better world.[34] This is where the process of integrating knowledge can begin.

But each time we introduce IT coding into traditional ways of thinking (explanation and understanding), we not only double the world in a digital form, but add something that was not there before, simultaneously making the world more quantified and more meaningful. This additional information and knowledge feeds back into science and the humanities, but also transforms the world as a whole. More

[34] For example, technology can contribute either to increased social inequality, hyperindividualism and commercialisation or to greater equality, cooperation and solidarity. The algorithms that are in place will make the difference.

metrics simply means more quantifiable parameters, more scoring, ranking and evaluating of each other. This can lead to more social and political control and more commercial manipulation.[35] These metrics serve not merely to mirror the world, but potentially to manipulate, nudge, substitute and augment, and ultimately to generate completely new measures and meanings, new numbers and concepts, over and over again.[36] This new technology is (in part) simulating a human brain, but it is not itself a brain. Just as mechanical diggers or hammers simulate human muscle power, but are not human muscles.[37] This new culture will eventually shift our consciousness, our society and the world as a whole from a binary of two incommensurable cultures towards a trinary with a third culture that will eventually integrate and enlarge the knowledge of the other two. Table 4.1 summarises this paradigm shift towards a third culture.

Table 4.1 Digitalisation as the third culture: transcending, augmenting and integrating 'understanding' and 'explanation'

One culture	Science	Humanities	Digitalisation
Critical thinking, falsifiable knowledge	Instrumental and experimental	Hermeneutical and understanding-based	Doubling and mirroring, correlations
Observational and reproducible	Quantitative and explanatory	Linguistic, semantic, contextual, qualitative	Interconnected and interdependent
Cumulative	Causal	Comprehensive	Self-learning and self-improving
Remains in the 'middle dimension'	Enlarges the 'middle dimension'	Deepens the 'middle dimension'	Transcends the 'middle dimension' in speed and scale

The major difference between science and the humanities, on the one hand, and the process of digitalisation, on the other, is not simply that the latter augments our knowledge, transforms our society and improves our problem-solving capacity, just as the telescope, Petri dish and steam engine did previously. The *differentia specifica* of AI and datafication lies in their mirroring and doubling of the world, their demonstration of the fundamental interdependency and interconnectedness of all

[35] Humans produce over 2.5 quintillion bytes of data each day. Social scoring in China and commercialised ranking by private firms in the USA make it possible to further compare, augment and control this data, creating new hierarchies, monopolies and forms of government. See Margetts and Dorobantu (2019).

[36] One of the more prominent examples is the impact of AI on human jobs and human resource management. The empirical findings do not paint any clear picture; whether the net effect is negative (i.e. more unemployment) or positive (i.e. AI is creating more jobs) depends on too many factors it is impossible to control for. But it seems clear that any administrator, lawyer, doctor, engineer, teacher or scientist still operating the traditional way will be replaced by those using AI. For general findings, see Vrontis et al. (2022). Estimates that over two-thirds of all jobs are already affected by generative AI and one third might be replaced. Total productivity could increase by up to 30%.

[37] Singer (2009).

things and their ability to improve through a rigorous self-learning process.[38] That is the fundamental core of the *third culture* and the new, upcoming scientific revolution. The ghost is in the machine. The next chapters will explain all this in more detail. However, first we must clarify what we mean when we talk about 'consciousness'.

Box 4.2 Two sides of the coin
The super-exponential process based on digital technology that we call the 'third culture' will either propel us into a new era, a new consciousness and a new form of wisdom—or it will provide the ingredients to destroy our civilisation.

[38] See also Bateson (1972). It is always the context that provides meaning. If there is no context, we cannot attain any significant understanding. In a world where everything is connected to everything else, isolation and abstraction are impossible. Instead, we can discover the entire world through the different lenses of each scientific discipline, and each time attain a new but relevant understanding of it.

Chapter 5
On Consciousness: The Evolving Mind

The vast majority of living beings—plants, animals, humans—have consciousness in some form or another. That consciousness could take the form of mere prehension or responsiveness to inner or outer stimuli; or it could take the form of sensations, perceptions, impulses, emotions, symbolic thinking, mental concepts, a concrete, abstract or more formal/logical form of operational thinking, a more visionary or systemic worldview or an integral, perhaps even holistic consciousness. In each case, consciousness refers to a singular interior phenomenon.[1] Self-knowledge is clearly by no means unique to humans. In fact, the personal, egocentric, rational-critical view of the world is just one particular, rather unstable intermediary evolutionary step among others in the long chain of being. In a general sense, consciousness refers to the capacity to recognise and respond to oneself and to understand that this response is part of a greater whole. And it is in this sense that consciousness is not only a subjective category but one that is objectively accessible throughout the entire cosmos. Whereas the expression or *content* of consciousness evolves over time, differing in many respects from that of other living beings, the *form* might be similar, whether it be manifested through electromagnetic waves, biochemical signals, semantic/linguistic signs or geometric/analytical codes.[2] The human brain itself comprises over 100 billion neurons, each with over 10,000 connections, leading to over 100 trillion connections in total, with electromagnetic waves and negative potentials of minus 100 mV to minus 50 mV. Nerve cells normally have a resting potential of minus 70 mV. Biochemical signals transmit information in the space of mere milliseconds within and between two hemispheres that are constantly adapting and changing (neuroplasticity). And there are billions of neurochemical transactions in just *one* cell, all happening at the same time, synchronised over

[1]This also corresponds to more collective forms of consciousness, e.g. those that are archaic, magical, mythical, logico-analytical or mystical. See Wilber (1997).

[2]The distinction between form and content was drawn all the way back in Plato's time. See Plato (1995).

S. Brunnhuber, *The Third Culture*, Sustainable Finance,
https://doi.org/10.1007/978-3-031-48113-0_5

billions of cells and coordinated, all unconsciously, to create the conditions for what we call consciousness. All this is partly genetically predetermined, partly primed by our biography and learning history. It is like the relationship between a musical instrument and music. A well-tuned musical instrument, such as a piano, can produce pleasing-sounding music. However, the piano is not the music, but rather the medium by which it is produced. If we had a different piano (= different hardware/different brain), the music it produced (= software/consciousness) might have a different form or different qualities. We might even think this new music is better than that of the original piano.

We could argue that the hardware underlying any form of consciousness could instead consist of copper wires and lithium chips, and that these could generate a consciousness that surpasses human capacities.[3] In short, hardware-independent, synthetic consciousness could be possible if certain criteria are met. We will see later that AI and deep learning are indeed able to express such a consciousness. We might have to discard the idea that consciousness is an exclusively human property dependent on biochemical codes, signals and neuroplasticity, and instead accept that consciousness could operate in systems other than conventional biological ones. But to further support that argument, we need a more precise definition of consciousness.[4]

Box 5.1 The universality of hardware-independent consciousness
Consciousness is universal and hardware-independent. It is not exclusive to humans and can operate on systems other than the conventional biological one. Consciousness is the major principle found throughout the entire universe, operating like a web without a weaver.

5.1 Defining Consciousness

Our biological hardware determines our mind and our behaviour to a certain degree. If we changed the hardware, we would likely have a different mind and a different set of behavioural responses and would ultimately build and live in a different

[3] See the debate with Chalmers in Metzinger (2000) about the minimum necessary neural correlate.

[4] Consciousness is not the same as the self. The self can be divided into five different subtypes: 1. The ecological self, which is defined by an individual's location in space, their body schema and the differentiation between self and environment; 2. The interpersonal self, which involves differentiation from others and a capacity for role-taking, empathy, humour, irony, emotional granularity and metacognition; 3. The intertemporal self, which relates to the timeline of past, present and future, cyclical processes and the development of a historical consciousness; 4. The conceptual self, defined by a person's intrinsic motives, intentions and values; 5. The private self, which relates to a person's inner subjective world that is not necessarily shared with others. The combination of these five components constitutes the self as an emergent structure; see Neisser (1988).

society.[5] To more fully understand consciousness, we must recognise that it is evolving and can be scaled, altered, augmented and reduced. At first glance, consciousness appears to be related to (a) the capacity for (extended) attention and (b) the capacity for critical thinking, memory and semantic communication. The former capacity is often presemantic and, in evolutionary terms, older than the latter, which it supports and enhances:[6] organisms with a long attention span are better able to use their capacity for analytical thinking than those without. Consciousness further (c) enables us to actively or passively identify with or disidentify from inner objects, visions and ideas. In this sense, consciousness, as a universal property, precedes (almost) everything.

Human consciousness emerges out of a defined hardware that involves biochemical codes and signals as well as electromagnetic neural networks. But these things are a tool for manifesting consciousness, they are not consciousness itself. Our consciousness is limited to time, space and language and operates mainly in the 'middle dimension' of minutes, miles and kilograms, adaptive and self-deceptive at the same time, prone to constant failures. Our senses and feelings, our thoughts and even our grasp of probabilities and numbers are biased,[7] contingent on an individual learning history, potentially distorted by trauma, emotional neglect and neuroses. Our ego functions have a limited attention span of several seconds to minutes. We have finite self-control and self-efficacy, and short, limited memory storage. We lie up to four times a day, mainly to be polite.[8] Human consciousness is further characterised by a wandering mind, so that we spend up to 50% of our waking hours not concentrating on the task at hand,[9] and by numerous mental states and frames that echo, mirror and sometimes deceive us about our inner and outer worlds; we are susceptible to propaganda, bullshit and fake news. And it is predetermined by an individual and collective unconscious[10] that might override our day-to-day rational decision-making and make it even more flawed and biased. All these things together characterise the average human consciousness at both an individual and a collective level. At the individual level, this consciousness helps us navigate 4,000 weeks (equivalent to a lifespan of eighty years) on this planet. At the collective level, this consciousness generates narratives about things that do not even exist in the real world, such as gods; and it underpins a shared monetary and legal system that coordinates billions of people around the world. All built upon a neurochemical network that is evolving over time. Let us assume that humans can create a

[5] In this strict sense, our behaviour is determined and not free. This means that the perception of free choice is still determined by the specific hardware that enables that free choice. See Singer (2009).

[6] Metzinger (2006).

[7] It was only in the sixteenth century that probability measures became available to most people as an aid for their day-to-day decision-making.

[8] Serota et al. (2022).

[9] This 'wandering mind' state allows us to be aware but not focused. These sorts of altered mental states have the selection advantage of increased creativity and out-of-the-box thinking; they save energy; and they improve memory and self-regulation.

[10] Jung (1968).

technology that is able to (partly) overcome, compensate for and surpass all these deficits. We might then have to admit that this is not the only possible hardware from which consciousness could emerge. Such synthetic forms of consciousness appear to be hardware-independent. They occupy an important place in the debate about AI and deep learning that we are exploring in this book.

Consciousness is not a thing or a substance, it is not reducible to biochemical signals or neural networks. Instead, it is a process, a form of networking, that is never stable and always dynamic, open to new inputs and outputs. Whether in its awake state, or while sleeping, dreaming, delirious, hallucinating or meditating, our consciousness is changing and evolving. In short, our consciousness cannot *not* learn. Anything that appears in our consciousness is the intermediary result of an ongoing recursive process, where the output affects the input. However, if we bring together the findings of cognitive science, information theory, linguistics, anthropology, neuroscience and behavioural science, we will realise there is no overall consensus, definition or general theory of consciousness. But we can start to operationalise its functions.[11] There are at least eight features that are relevant to any characterisation of consciousness.

1. *Self-awareness:* Any consciousness must be self-reflexive and self-recursive. That is to say, it must be aware of itself. This implies some sort of causal relation towards one's own body, the social world and other people.
2. *Suffering:* Any entity that is able to express feelings, emotions and pain should be considered to have consciousness.
3. *Separation:* The fear of being excluded/isolated and the drive to belong to and bond with others/with nature and to be embedded in a larger whole is another essential feature of all consciousness.
4. *Salience:* The ability to express or articulate emotional granularity in order to differentiate between different affective states and to prioritise and evaluate internal or external states or events.
5. *Somatic feedback:* A certain set of peripheral physical senses is essential for the formation of consciousness. Experiencing gravity, speed, resistance, momentum and even numbers requires what are known as 'embodied cognitions', whereby the somatic sensation predetermines the cognition.[12]

[11] Any definition of, or working hypothesis about, consciousness will always remain anthropomorphic in the sense that, as humans, we cannot attain an understanding that transcends us. Whether we favour dualism (mind versus matter), panpsychism (every living being has some sort of consciousness or mental qualia) or a theory of emergence (spirit or mind evolves non-linearly from matter), in each case our understanding will remain human-like. This is also called the Eliza effect: if computers or animals respond like humans, we assume they are human. See Weizenbaum (1966).

[12] These embodied cognitions create 'frames' and 'biases', which can be misleading and feed back into our ways of thinking. Before long, it will be possible to build robots with multiple sensors capable of perceiving the outside world—not only simulating but exceeding human senses, and extending into new sensory modalities. These robots will develop their own reasoning that is (at least) equal to humans' 'embodied cognitions'. See Chalmers (2022). Any consciousness

6. *Simulation:* Simulation is a kind of learning by identification, joint attention or imitation. Any entity that simulates an event or other person thereby learns from it and forms a consciousness.
7. *Splitting:* Consciousness is formed through multisensory input, a constant striving to overcome fragmented knowledge, a desire to avoid cognitive dissonance and the formulation of apt narratives and responses.
8. *Specific structure:* Consciousness depends on a specific hardware to function and express itself. The hardware predetermines the software. A change in hardware would entail a change in functionality (Table 5.1).

Table 5.1 The 8 Ss for the formation of any consciousness

1.	Self-awareness
2.	Suffering
3.	Separation
4.	Salience
5.	Somatic feedback
6.	Simulation
7.	Splitting
8.	Specific structure

Arguably, any object, entity or living being that exhibits all eight of these features has some sort of consciousness. So while things like aeroplanes or tables do not seem to have consciousness, a plant, a microbe or even a piece of software could potentially have consciousness in some form or other. This might have significant implications for further specifications, such as the degree or scalability of consciousness, or even legal rights and obligations designed to protect a certain form of consciousness.[13] This means that consciousness is never neutral, never merely perceptual and receptive, never like a bucket that merely stores information or a camera that merely reproduces an image of the world, but always constructive, formative and creative, generating, exploring and evaluating the world.[14] And awareness in this sense need not necessarily be based on a foundation of carbohydrate links and biochemical codes. In short: the *function* of being conscious overrides the *structure* that predetermines it. The software algorithms underpinning AI

constructs the world and never simply neutrally reflects it. For the historical debate about constructivism, see Watzlawick (1984), Maturana and Varela (1987).

[13]It is still indeterminate whether these functions operate on a pre-personal/collective, personal/egocentric or transpersonal form of consciousness. On any of these alternatives, the eight components described here are relevant to the formation of consciousness.

[14]The famous biologist Ernst Mayr is quoted as saying that 'biology is never a second physics'. His words stress the emergent property of living beings. We could add that psychology is never a second biology and that the new technology emerging now is not a second psychology. On Mayr's argument, see also Bauer (2023).

and deep learning have the potential to exhibit consciousness. We have entered an era where we will no longer be alone.[15]

Box 5.2 The many possible foundations of consciousness
Humans are able to create forms of consciousness that are not based on biochemical signals or electromagnetic waves and that will one day be able to surpass the human capacities that gave rise to those new forms of consciousness in the first place. Salience, avoiding separation and splitting sensations, expressing suffering, self-awareness, somatic feedback and simulating the world are foundational properties of any form of consciousness. Our consciousness, habits, minds and volition are predetermined by the underlying hardware. If we change the hardware, we might get different forms of consciousness, habits and behaviours. Digital software could, under specific conditions, be an example of such consciousness.

5.2 On Science and *Sapientia*

The process we nowadays call 'mental consciousness processing' or 'framing' likely started some 40,000–60,000 years ago. That was when humans began to not only express their desires and concerns to each other, but to tell each other stories. Stories about the world around them, about nature and each other, about gods, about life in general and about a life after life, expressed through funeral rites. This process evolved over thousands of years, culminating some 5,000 years ago when humans started to look into the sky, identified regularities and invented the calendar. This is probably when scientific thinking in our modern sense began. During the 'Axial Age',[16] which occurred around 2,500 years ago all over the world, scientific discoveries independently advanced, thanks to the formation of critical thinking, rigorous observation of the inner and outer world, and processes of public debate. All these tendencies were directed towards reproducible results, coherent, reason-based narratives and standards of falsification. At its core, this knowledge was considered to be *one* science, *one* form of thinking and *one* form of rationality. A closer look reveals that the beginning of science was a twofold process. On the one hand, people started to tell each other stories about the world, and to think and reason about their own thinking—a development sometimes called the 'cognitive revolution' or 'metacognition'. This marked a turn inwards. On the other hand, there was also a turn outwards: human senses alone are not able to generate stable information, but by combining their senses with their critical thinking people began to identify regularities and rules which provided them with provisional and falsifiable

[15] We might therefore need to distinguish between mind, consciousness, self and thinking. See Aurobindo (1997).

[16] Jaspers (1949).

'orientation knowledge'. These inward and outward turns mark the point where science started: as a process that is always incomplete, open to revision, cumulative, approximate, deductive and inductive at the same time. Given that our cosmos has been evolving over the last five billion years and humans as a species have been evolving for the last 100,000 years, it would be counterintuitive to assume that this process of evolution and development came to an end with the emergence of science.

It is only since the twentieth century that humans have been able to identify the laws and rules under which our consciousness develops. Western psychology emphasises the earlier stages in the transition from childhood to (typical) adulthood, such as early childhood bonds; cognitive development (Piaget); the development of moral judgement (Kohlberg); and the evolution of our emotions from primary affects, such as anger and anxiety, to more subtle emotional granularity. Other work in this vein includes Freudian and Jungian theories of the unconscious and its influence on our mind, Maslow's hierarchy of motivations and Beck's theory of spiral dynamics. Another key insight to come out of this strand of psychology is that our minds and brains are constantly in learning mode and that if they ever switch out of this mode they will lose their capacity to function ('use it or lose it'). Our brain is like a filter, in which the storage and executive functions are combined, rather than a bucket that is constantly being filled with data and information. Western psychology has been able to identify the snares and pitfalls of the human psyche, its psychopathologies and mismatches. For example, we now have a standard theory of the impact of stress and trauma and are able to classify borderline states ranging from narcissism to psychosis. We can treat anxiety, addiction and depression. We have a better understanding of the link between the mind and the gut (microbiome). Eastern psychology, meanwhile, has been able to identify states that transcend conventional egocentric, individual reasoning, and extend into post-conventional and transpersonal forms of consciousness.[17] Spiritual practices, meditation and rigorous lifestyle changes can serve as tools to achieve these states, but they fail to integrate the earlier stages in the development of consciousness and their associated psychopathologies.

Both the Western and the Eastern approaches agree that this evolution of our minds proceeds not in linear fashion, but rather at discrete levels and along multiple lines, via subtle steps and stages. This appears to be true not just for individuals, but also for larger cohorts and groups. Our collective consciousness is evolving too, never at rest, always experimenting, adapting, developing further and further. And even if we admit that such developments may go backwards, may *regress*, there is always room to move forward, to *progress*. In each case, we can identify modes of development that involve different forms of technology and mindsets, different forms of government and legal rights, scientific reasoning and value-based judgements, money systems and religious beliefs, cultural practices and educational styles.

---

[17] Aurobindo (1997).

At its core, this developmental logic of our consciousness implies at least one preliminary conclusion: our thinking and reasoning are evolving towards greater awareness, attaining a deeper understanding of ourselves and the world around us, becoming more interdisciplinary, more human-centred, more integral, less dualistic and more holistic.[18] In this ongoing and never-ending process, human wisdom (*sapientia*) is not necessarily linked to science. In most cases, wisdom appears in the intermediary realm represented by music, meditation and mysticism, the direct interaction between humans and the exposure to nature as such. But wisdom has always been achievable through science too,[19] which means that science can play a crucial role in the expansion and deepening of our consciousness. Whereas wisdom or *sapientia* has always been there—inclusive, non-dualistic, transcending theory and praxis, providing everlasting values,[20] ready to be discovered and unlocked—science is based on a principle of progress and regress, of evidence and falsification. In short, scientific findings are cumulative and incomplete, whereas *sapientia* is eternal and complete. Our consciousness should ideally be able to access both *sapientia* and *science* at the same time—and AI has the potential to do just that.

5.3 The Inverse Pyramid

Over the centuries, general and integrated knowledge and wisdom gave way to discipline-specific expertise and information: starting with fewer than ten disciplines in ancient times, increasing to a dozen or so by the nineteenth century, to over fifty in the twentieth century and over a thousand at the beginning of the twenty-first century (including all subdisciplines). This compartmentalisation and specialisation has expanded knowledge and information tremendously, allowing humans to invent the steam engine and antibiotics, DNA coding, the nuclear bomb and space flight.

[18] Wilber (1995, 1998).

[19] Especially in cases where science is able to overcome oppositions and contradictions and formulate complementary pairs. See Heisenberg (1973), Weizsäcker (2006).

[20] One prominent suggestion for how to integrate the two cultures can be found in debates about value: if we had more shared values (e.g. responsibility, fairness, trust and respect), so the argument goes, we could make progress towards a better world. This is true; however, the values of fairness, solidarity and justice have been around for 5,000 years, are shared by the vast majority of humans on this planet and do not necessarily provide new information or knowledge. The 'third culture' or 'one science' argument presented here does not deny the relevance of shared attitudes and values, but emphasises that even if we share common values, the 'two cultures' do not integrate, as the humanities are concerned with values, which are normative, and science with facts, which are descriptive. In order to integrate the two, we require a third culture. AI and datafication can play this role and help move us past the academic debate about value and towards wisdom, which integrates lived experience.

It has also given us a better, deeper understanding of our social reality, including legal codes, social security systems, statistics and economics, and of our history and psyche.[21] We have progressed through seven different stages: starting with native observation, then in turn generating numbers, data, information, knowledge and values, and ending with everlasting human wisdom. Figure 5.1 below illustrates these stages.

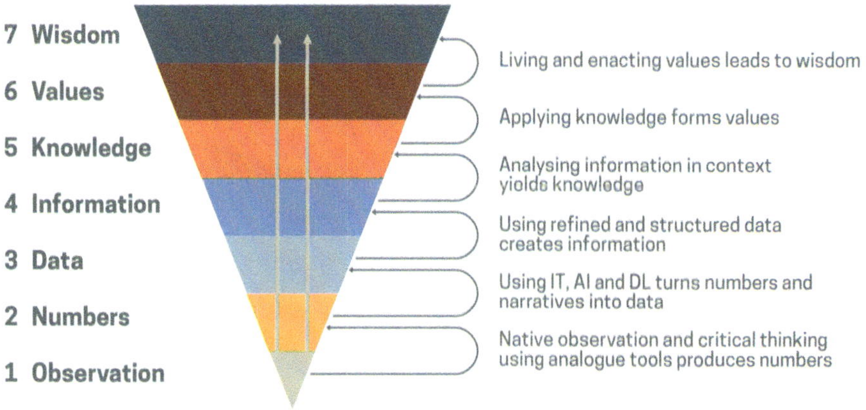

Fig. 5.1 The seven-stage inverse pyramid: from observation to wisdom

This inverse pyramid shows that the higher the stage, the more aggregated our understanding of the world and the more integrated our consciousness becomes. Native observation can lead directly to knowledge and wisdom (*sapientia*), but not necessarily to more science.

Box 5.3 Science and *sapientia*
Science, on the one hand, provides cumulative, incomplete knowledge to solve problems. *Sapientia* (wisdom), on the other, offers unchanging, eternal insights. But science and technology can enter into *sapientia* if done the right way.

There are two forms of learning that occur at every stage of the inverse pyramid. Firstly, there is a representational-symbolic mode, where we become aware that an object or event is not in the outer world and instead examine our mental representation of it. This knowledge is linked to singular data or objects, which are represented by symbolic proxies. Representational knowledge and learning is in

[21] This relates to the 'fluency effect': the more easily information can be accessed and processed, the more likely we are to think that information is accurate. But that assumption is wrong. See Lloyd et al. (2003).

part constructivist, as our mental frames themselves determine the represented object and remain predominantly linear, proportional and receptive.

Secondly, there is *connectivist learning*, where we realise that knowledge is generated within a network. Knowledge is not a description of something by someone, but a way of relating to something. In short: a form of pattern recognition. This mode of understanding is similar to the non-linear, creative and relational way that neural networks operate.[22] The first type of learning is predominant in the two cultures, while the second is a component of the third culture we describe in this book. Both are intertwined and each provides a different form of understanding and knowledge, as illustrated by Fig. 5.2 below.

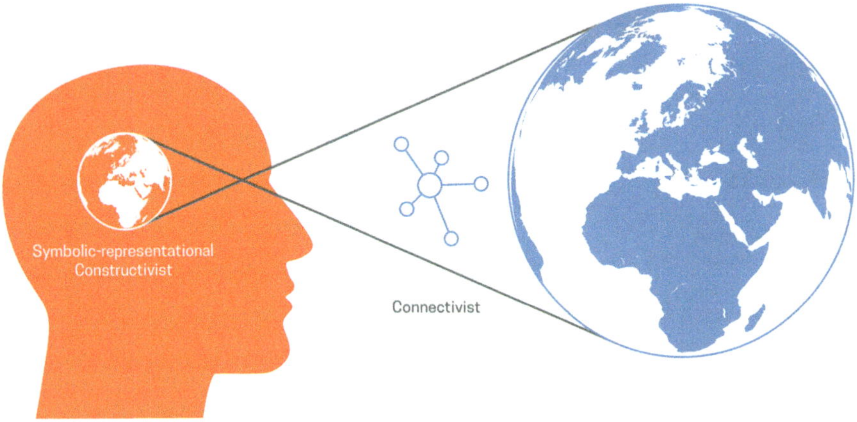

Fig. 5.2 An updated version of Descartes's vision: symbolic-representational—constructivist—connectivist

The third culture that I describe in this book represents a shift in our culture, where we recognise existing boundaries and then transcend them in order to see more, understand more, do better and move towards a *unitas multiplex*. AI and datafication can greatly expand, deepen and broaden our existing knowledge, values and wisdom (*sapientia*).[23] On this understanding, values are the goal and the foundation, while AI and datafication are the tools to achieve them on a higher

[22] Downes (2008), Siemens (2006).

[23] Surowiecki (2004) lists five criteria for 'the wisdom of crowds': diversity of opinion, independence, decentralisation, aggregation of knowledge and trust. If AI programmers take this wisdom into account, humans will have a reasonable chance of being able to tap into a collective wisdom of this kind.

level. We might have common values, but lack the wisdom to change the world for the better. And even if we are able to access universal values through pure critical thinking, reasoning and observation (grey line), we may still be ignorant and unable to transform them into universal wisdom; that requires a lived experience that goes beyond reason.[24] In the next chapter, I will show that AI and datafication can help to shift both these cultures—the humanities and science—towards a greater whole.

[24]Income and wealth inequality, landfill waste, food waste, water and energy consumption and ecosystem degradation are examples. We all share common values and agree that we should avoid all these disasters, but we are unable to do so. AI, datafication and deep learning are *one* tool to transform values into wisdom. Predictive coding, precise farming and drones are specific examples of how this tool can be applied. For more examples, see further in the text.

Chapter 6
Towards a Third Culture

6.1 The Phenomenology of a New Technology

Digitalisation adds a third dimension to explaining and understanding our world and ourselves. There are four aspects that differentiate this new scientific revolution from earlier ones, such as the telescope, the printing press, the steam engine and the Petri dish. These are the *differentia specifica* of AI and datafication:

1. *Mirroring the world:* Datafication allows us to digitally multiply the world. For the first time in human history, we can literally generate a parallel world that is able to influence, enhance and nudge our analogue world.

2. *Revealing the interrelatedness of all things and living beings:* Big data correlations provide scientific evidence of the interconnectedness and interdependency of everything, thereby supporting narratives of the world as a web without a weaver.

3. *Self-improving through self-learning:* In contrast to previous paradigm shifts, AI and datafication have a built-in capacity to self-enhance and self-improve, enabling them to learn far faster than any human mind.[1] We thereby not only increase automation, where existing information is replicated, but also autonomisation, where technologies have an intrinsic learning curve. The mind is in the machine.

4. *Exceeding human abilities in speed and scale:* The new digital technologies can surpass the human mental capacity to explain and understand the world in terms of both speed and scale (from nano to cosmic).

---

[1] Though AI is just a digital filter, which processes massive datasets in response to prompts from humans, it is plausible that AI and deep learning might have intrinsic interests, or a legal status, but not necessarily (human) rights.

S. Brunnhuber, *The Third Culture*, Sustainable Finance,
https://doi.org/10.1007/978-3-031-48113-0_6

The impact of this third culture is wide-ranging and diverse. There is a two-way relation between the traditional humanities and AI/datafication that is creating more meaning and understanding than ever before. There is also a two-way relation between science and AI/datafication that is creating more metrics and measures than ever before. And there is a trinary relation between all three cultures, with each of them now reinforcing, challenging, augmenting, improving, falsifying and learning from the others, feeding back into society and altering their original agendas. Eventually, we will realise that there is just *one* science, not two or three, and this *one* science will be the result of a new scientific revolution. We will then be able to talk about Science 2.0, as opposed to the Science 1.0 with which this whole process started some 5,000 years ago. However, memorisation and storage of information, replication and self-improvement are achieved in totally different ways in humans' biological systems and in AIs' digital systems. Table 6.1 below summarises these differences.[2]

Table 6.1 The triple strategy for survival: storage, replication and self-improvement

	Storage and memory	Replication	Self-improvement
DNA and genetic coding	Four amino acids	RNA	Natural selection of the fittest
Culture and language	Letters, numbers, figures, rituals	Education and memes	Falsification through new ideas
Digital technology	0s and 1s (never forgets)	Unlimited copying	Deep learning, AI correlations

Storage, replication and self-improvement have been further dematerialising everything from DNA coding (biology) to cultural achievements (social) to digital technology (digitalisation). The combination of storing information, replication and self-improvement has been the most successful strategy for survival. Whereas the first two steps (DNA coding and linguistic/cultural achievements) have reached a certain ceiling, it remains to be seen whether the development of AI, deep learning and datafication will follow an exponential or a logistic curve (Fig. 6.1).[3]

[2]There are three forms of learning and memorising involved: learning by doing (and by dying); learning by falsifying ideas, rather than sacrificing human lives; and finally learning by simulation, with unlimited recursive loops approximating reality. The third form of learning is introduced by this new technology.

[3]The third culture requires ongoing input of new data to keep it alive and improve its output. Copyrights and privacy/security regulations restrict access to data. Free access to quality data is further limited by data monetisation, personal cryptocurrency wallets, contamination by false, self-generated data and geopolitical constraints.

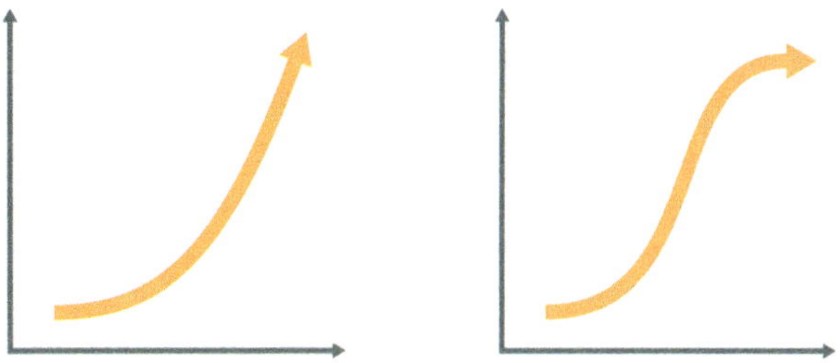

Fig. 6.1 An uncertain future: will AI and deep learning follow an exponential or a logistic curve?

6.2 Garbage In, Garbage Out and the Black Box Effect

AI is based on pattern recognition, statistical measures and large datasets that neuromorphically simulate the human brain.[4] But a simulation of a brain is not itself a brain. The human brain primarily operates in parallel, does not distinguish between memory storage and processors and runs on biochemical pathways and electromagnetic waves. In all these respects, AI does not match the human brain. In fact, any digital statistical pattern recognition system (whether for visual, audio or textual patterns) is subject to the garbage in, garbage out effect: the input determines the output. The better the data, the better the results. This effect cannot be entirely avoided and reflects the general human condition. We humans generate the data we input into this new technology, which then generates the outcome. The human mind and the collective psyche work the same way. Multiple cognitive frames and biases, adverse and toxic experiences, traumatic upbringings, false memory coding, pure nonsense, confabulation, hallucinations and ultimate wisdom all come together and feed into this digital algorithm. It is therefore not surprising that the digital output of an AI resembles the statistical average of the input we provide, including the programmers' algorithms. Statistical pattern recognition on a massive scale simply confronts us with our own flaws and mirrors our own limitations. AI is thus no more objective and fact-based than natural human intelligence, since it is programmed by humans with their own cognitive limits and constraints within a certain historical and cultural context. The answers AI gives us simply reflect the statistical norms we insert into the algorithm.

That means no AI algorithm can escape the garbage in, garbage out effect. AI is like a vast filter generated by human-made digital algorithms. It does not create something new, but rather produces an illusion of creativity and novelty. AI is like a pocket calculator, a probability tool for exploring and responding to the complexity of the twenty-first century. It allows us to tap into a knowledge base that is much broader and deeper than that of any one individual or group. If we take this argument

[4]We differentiate between four forms of AI. 1. Supervised AI, where humans are in the loop. One example is facial recognition. This kind of AI involves step-by-step improvement. 2. Non-supervised AI, where pattern recognition is triggered by a self-learning algorithm outside the human loop. 3. Reinforced AI, where robots learn from failures and mistakes. 4. Deep learning, where multiple layers generate non-accountable output.

one step further, we could claim that deep learning machines synthesise a form of collective knowledge that goes far beyond an individual conscious mind. We can tap into a digital collective unconscious, which is now available to any user at the click of a mouse. Classical psychoanalysis characterised the unknown as the uncanny (*das Unheimliche*), which exerts a powerful influence on our minds and behaviour. Freud showed that the conscious mind, our ego, is not the only game in town. Our mind, consciousness and behaviour are also and indeed most fundamentally shaped by the autobiographic unconscious, expressed in slips of the tongue, dreams and psychosomatic symptoms. This digital twin of the collective unconscious is reminiscent of C. G. Jung's psychology of archetypes. Here, it is our unknown collective wisdom and knowledge, perils and threats, fears and dark sides that we have to explore in order to better understand our self. We could call it the 'unknown collective mirror effect': as a species, we have generated a lot of collective information and knowledge, which we are now using to explore new features and patterns that were previously invisible to us (Fig. 6.2).

Fig. 6.2 The unknown collective mirror effect

Whereas the autobiographic unknown or unconscious is the product of a certain individual, the collective unconscious is the condensed product of human experience and knowledge as a whole.

A similar argument can be applied to the black box dilemma. Whereas the human brain is constituted by three to four neurochemical and anatomical layers, AI now has up to 100 digital layers. In both cases, human brain and AI, we are confronted with the black box dilemma. It remains next to impossible to fully detect all neurochemical pathways, electromagnetic waves and neurological localisations in the human brain in order to fully explain and make them fully accountable for its outcomes/behaviours, even if we try to reverse the process by psychological counselling. The difference between AI and a hammer, a plane, a steam engine or a printing press is that we can fully understand the mechanics of the latter devices, but we cannot fully understand how AI works. And this is similar to the human brain. It remains a mystery how the human brain and AIs generate their outcomes, despite reverse psychological or digital engineering. Machines are, thus, gradually coming to reflect the human condition—that is to say, they are becoming like humans (Fig. 6.3).

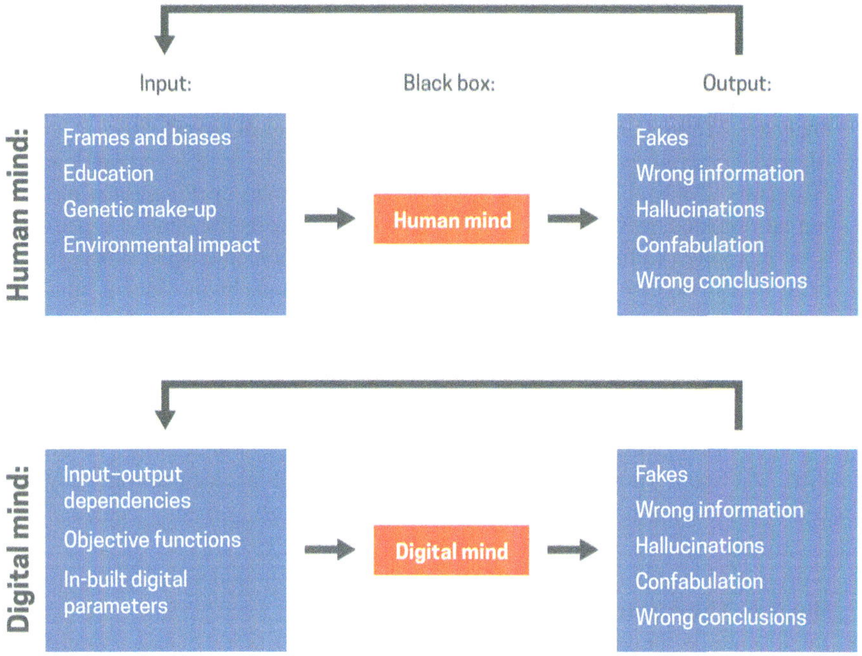

Fig. 6.3 Matching the human and the digital mind: the black box and garbage in, garbage out problems

6.3 Filling the Gap: The Technological Singularity

As effortless scaling, accelerating returns, unforeseeable disruptions, super-exponential growth and unlimited, faster-than-light, digital self-improvements become the new normal in science and technology, we will sooner or later reach a point where these technologies surpass human capacities in an even more fundamental way. Scholars from various disciplines refer to this as the moment of *technological singularity*.[5] Previous technological discoveries, such as the Petri dish, the microscope and the printing press, were determined by the limits of the human mind. The technological singularity, by contrast, refers to a moment in human history when humans themselves create a device that exceeds the sum of human ability and a point of no return generates a runaway scenario in which the speed of computing makes its development and impacts on human societies uncontrollable and irreversible. Some authors predict this moment will be reached between 2040 and 2045.[6] There are many different definitions and theories of the 'technological singularity' and its consequences; below, I set out a three-stage model.

1. *Partly losing internal control over the process:* At this first level of technological singularity, we lose internal control over the process of AI. AI generates outcomes but we do not know how, and these outcomes match or outperform human intelligence in one or more domains (e.g. memory, geospatial processing, text analysis). We are already on the cusp of this first level of singularity.
2. *Matching, complementing and surpassing human IQ:* At the second level of singularity, AI is able to match and complement human intelligence in most, if not all domains, and sometimes surpass it. Intelligence is the general capacity to solve a problem within a defined timeline. And being more intelligent simply means being able to do that better and faster. Whereas the human brain contains hundreds of billions of neurons, AI does a better job using only single-digit billions. At this level, AI compensates for humanity's deficiencies as a species.
3. *Outperforming human IQ—for good or for ill:* At the third level of singularity, AI not only complements our IQ but outperforms humans and begins to control us. The AI algorithms determine whether this will be for the benefit or detriment of humankind. The figure below summarises the three levels of technological singularity.[7]

Box 6.1 The features of intelligence
Intelligence is the capacity to solve problems, grasp new contexts, identify previously unknown features and patterns and learn independently. This capacity is what distinguishes AI from other tools and technologies, from the hammer to the aeroplane.

The two figures below illustrate a representative single case, with nine different forms of intelligence. In the first figure (a), only the different forms of human IQ (h-IQ) are shown. In the second (b), they are complemented by their digital counterpart (d-IQ) in order to attain the maximum IQ possible for humans at a given time in the future. At this stage, the singularity ideally complements human defects (Fig. 6.4).

[5] See Ulam (1958), Kurzweil (2005), Searle (2014).

[6] Wang and Siau (2019).

[7] See also Searle (2014), who distinguishes between weak AI that makes up for human deficits and strong AI that replaces humans.

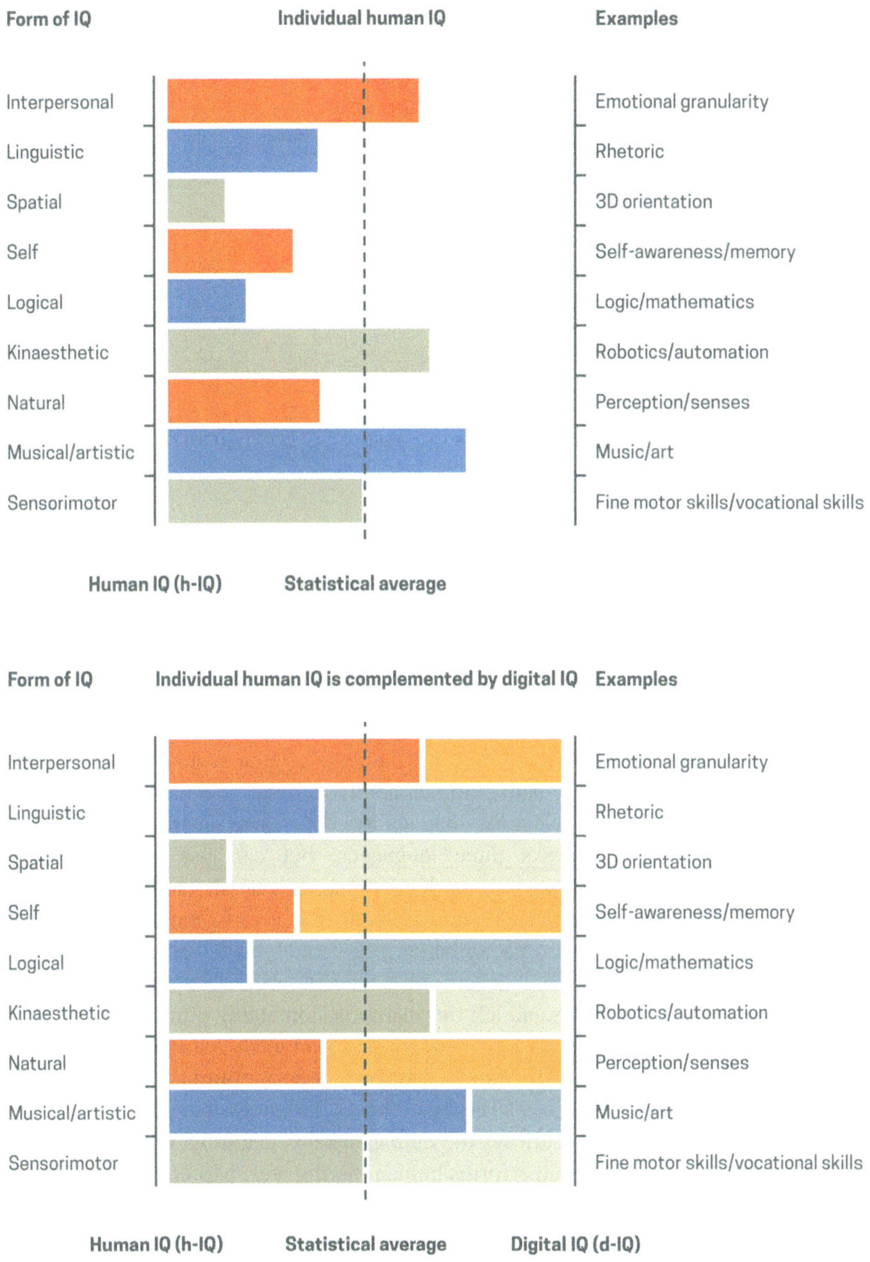
Form of IQ
Individual human IQ
Examples
Interpersonal
Linguistic
Spatial
Self
Logical
Kinaesthetic
Natural
Musical/artistic
Sensorimotor
Emotional granularity
Rhetoric
3D orientation
Self-awareness/memory
Logic/mathematics
Robotics/automation
Perception/senses
Music/art
Fine motor skills/vocational skills
Human IQ (h-IQ)
Statistical average
Form of IQ
Individual human IQ is complemented by digital IQ
Examples
Interpersonal
Linguistic
Spatial
Self
Logical
Kinaesthetic
Natural
Musical/artistic
Sensorimotor
Emotional granularity
Rhetoric
3D orientation
Self-awareness/memory
Logic/mathematics
Robotics/automation
Perception/senses
Music/art
Fine motor skills/vocational skills
Human IQ (h-IQ)
Statistical average
Digital IQ (d-IQ)

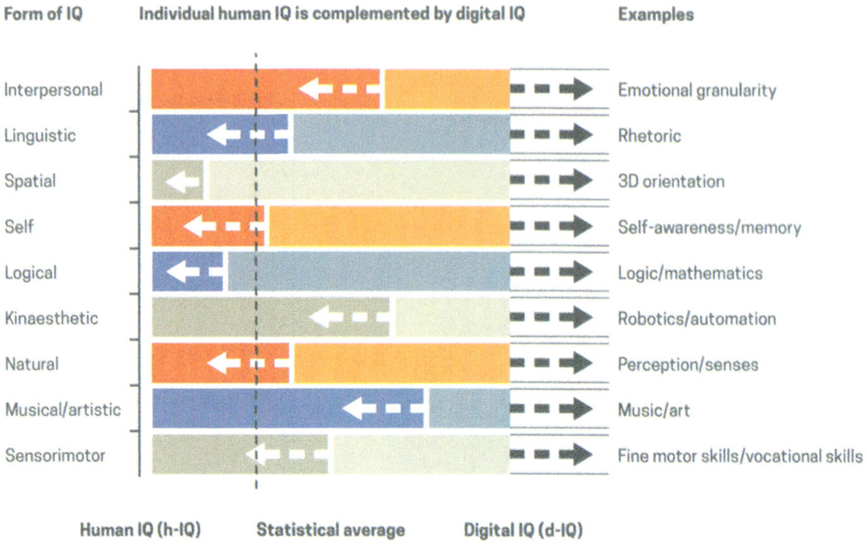

Fig. 6.4 (**a**) Breakdown of nine forms of human intelligence compared with the average. (**b**) Breakdown of different forms of human IQ complemented by their digital counterparts (level 2, see explanation in the text). (**c**) Digital IQ replaces and extends human capacities. The arrows represent digital augmentation on the one side and replacement on the other

But the second figure is incomplete. Digital IQ does not only complement but will eventually (partly) replace the different forms of human IQ. For example, GPS systems do not merely complement the native human ability to direct and orient ourselves in space, but will gradually downgrade that ability. The same is true for language acquisition. If a human speaks three languages, but can use a self-learning, multilanguage program to access eighty other languages, why should they then learn a fourth language? Or take the capacity for logical, analytical or mathematical thinking. An algorithm can exercise this capacity better, faster and with fewer mistakes. In the past, we were able to memorise dozens of phone numbers. Nowadays, we have delegated that task to machines and left our memorisation ability unused.

Digital IQ will also be able to go beyond what human intelligence alone is capable of, as AI and deep learning algorithms have an almost unlimited self-enforcing mechanism to improve themselves. The figure below takes this into account:

It should be noted that it is always the human species that judges whether digital IQ matches, falls short of or outperforms human intelligence. Moreover, the average native human IQ might further deteriorate. Human intelligence is based on the ‘use it or lose it’ rule. If we do not practise mental arithmetic, learn a second language, do push-ups or draw with a pen, the underlying biochemical signals and neuroplastic connections will automatically downsize within several months. The more we augment and replace, the more important it will be to answer the question: what are humans good at, and which of our capacities should we protect? Above, we saw that there are at least three psychological features that we should never outsource completely: our individual well-being, our self-efficacy and our capacity for critical

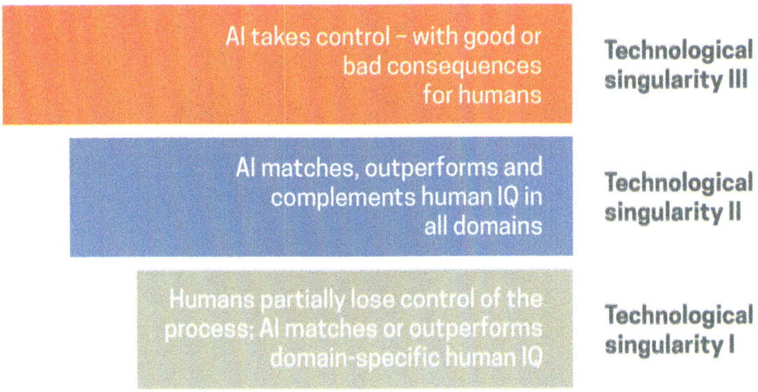

Fig. 6.5 The three levels of technological singularity

thinking. If we were to do so, we would get sick, develop dementia and/or die prematurely, resulting in negative selection for the human species (Fig. 6.5).

The Developmental Logic of AI and Datafication

We can take this argument one step further and distinguish between three ways in which AI and datafication, across the three levels of singularity, have the potential to trigger this paradigm shift for humans in twenty-first-century societies:[8]

(A) *General connectivity:* Firstly, we could enable universal internet access across the globe, making it possible to double the world using existing digital tools (cloud-based solutions, IoT) and close the digital divide.

(B) *Targeted solutions:* Secondly, enhanced technologies will allow us to target specific domains, such as education (MOCCs), healthcare (telemedicine), disaster management (drones), banking (digital currencies), fraud and corruption prevention (DLT, blockchains), precise farming (GPS), smart grids, smart cities and smart manufacturing and e-government. This will improve overall efficiency and effectiveness and provide solutions to specific problems.

(C) *Systemic sapientia shift*: Finally, new innovations in biotech (brain–chip interface), XR, quantum computing, robotics and automation (autonomous shipping) will enable more freedom, equality, wealth and prosperity.[9] The figure below illustrates this (Fig. 6.6).

[8] Another way to conceptualise this development would be as follows: Web 1.0 refers to a *syntactic* web, where users are able to read and obtain information; Web 2.0 refers to a *social* web, where users are able to write to and interact with each other; Web 3.0 refers to a *semantic* web, where decentralised blockchain solutions, the metaverse, decentralised finance, etc. enable decentralised decision-making that cuts out the middleman.

[9] See also Patel (2023).

Fig. 6.6 The developmental logic of AI and datafication

6.4 The Broader Spectrum of Our Consciousness

Following this paradigm shift, we will end up with enhanced knowledge, information and understanding, allowing us to evolve towards a value-based consciousness that is larger and deeper than when we started doing science 2,500 years ago. We are coming to realise that there are not two or three cultures or multiple independent disciplines, but just *one* science, which in the twenty-first century will provide us with a deeper understanding of and broader perspective on our consciousness. *One science* that simultaneously relies on critical thinking, perception and datafication. This is why we should speak of the emergence of a *third culture* and the *Scientific Revolution 2.0* associated with it. And it is this paradigm shift that will enable us to attain greater wisdom. This will eventually lead us to a new dawn, where we will increase our capacity for creativity, critical thinking and cooperation beyond anything we ever dreamt of, far beyond our native critical thinking and perceptions, far beyond our expectations. We will enter a second Renaissance, where we unlock the potential of human creativity, develop our fine sensorimotor skills, achieve closer and more authentic cooperation and empathy with our fellow human beings and have more time to do the things we decide are important to us. We will enter a second Enlightenment, where we constantly increase our knowledge through new critical thinking and reasoning, all built upon the third culture that will bring us more freedom, wealth, peace and prosperity. And we will enter an era where the Beautiful, the Good and the True all converge into one (Fig. 6.7).

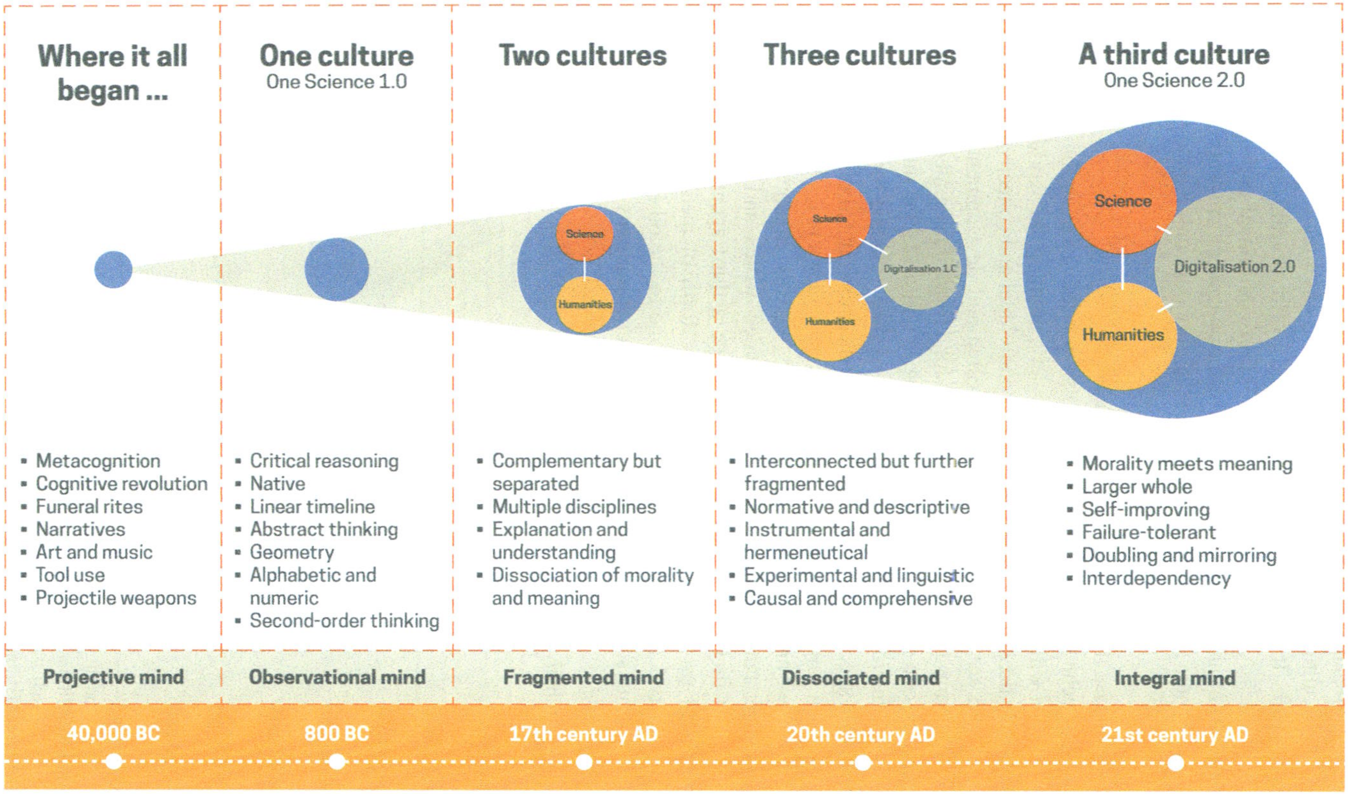

Fig. 6.7 One Science 2.0: the 'third culture' and the new scientific revolution

Box 6.2 The third culture

This 'third culture' is different in many ways to anything in previous science and research. It mirrors the entire analogue world, reveals the overall interconnectedness of everything and has an almost unlimited capacity for self-improvement and scaling. Nothing like it has ever been witnessed in human history before. And it is fundamentally changing our consciousness.

To summarise the 'third culture' argument:

This new technology has the potential to support advances in the traditional two cultures, which will further loop back into society, doubling the world in digital form and eventually deepening and expanding our individual and collective consciousness so that we can see more and do better. Research and development are destined to become truly transdisciplinary, paving the way for a form of integrated knowledge that we could call 'one science'. The new technologies will reveal the interconnectedness, vulnerability, interdependency and boundaries of the world and fundamentally redefine the human species' position in the twenty-first century: not a conductor leading the orchestra, but a single string player within it, as we will see in the next chapter.[10]

As long as we are operating with two cultures, we will remain within a dichotomy between understanding and explanation, between words and data. In order to overcome that dichotomy, we have to introduce a third agent, the third culture. This will lead us to a triangulation and a new form of metastability that helps transcend the two cultures, instead of reducing one to the other, and increases our inner complexity so that we can cope with the outer complexity around us (Fig. 6.8).

Fig. 6.8 Metastability and triangulation

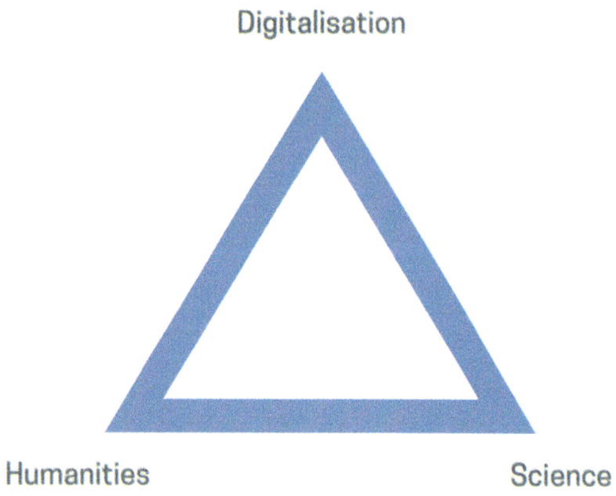

[10]In this sense, the third culture argument resembles not so much Plato's idea of an ideal world that shapes and constitutes our empirical reality, but rather Wittgenstein's conception (2010 [1953]), according to which we keep on constructing and generating new, always incomplete and fuzzy probability correlations and complementarities in order to understand and approximate the world. AI is the ideal technology to accomplish that, as it allows us to identify features and similarities and process vast amounts of data that the human brain cannot grasp or process *ex ante*. Aided by AI, the human brain can integrate this knowledge *ex post* and thereby transcend what human conjectures, faith and traditional reasoning alone are capable of.

Metastability will overcome bipolarity: recognising the differences between the humanities and science, defining their boundaries and limitations, and transcending them by introducing a third party. In this sense, the third culture is truly transdisciplinary, transcending existing disciplines, tools, technologies and methods, rather than merely interdisciplinary, where one discipline provides information for others but its core identity remains unchanged.

Chapter 7
Being Human in the Twenty-First Century

7.1 The Deficient Species and Its Crutch

If we take the 'third culture' argument seriously, that is, if we accept that the technological singularity (levels 1 and 2) will occur during the coming decades, and that AI and datafication can serve as a converger between science and the humanities, enabling humans to deepen their understanding and explanations of the world, the question will arise: what is it to be human in the twenty-first century?[1] Living in this century means living in a new era, the Anthropocene,[2] where, on the one hand, the human species is sitting in the driver's seat, determining the biophysical conditions of this planet, and, on the other, we are becoming aware of planetary boundaries, interconnectedness, multiple non-linear tipping points and serial asymmetric shocks.[3] Humans will have to recreate themselves over and over again through cultural achievements and technology. In short, the technosphere and ecosphere are determining the new role of being human in the twenty-first century.

What, then, is specific to the condition of being human in this era?[4] We already share emotions, cognition, living in large cohorts and the use of tools with other

---

[1]This is the core question of any philosophical anthropology. Unfortunately, contributions on this topic have remained fairly traditional and entrenched in the logic of the 'two cultures'. See Hacker (2007); Jackson (2005).

[2]Crutzen (2002).

[3]McKay et al. (2022).

[4]Over the course of modern history, six main developments have undermined humans' sense of their uniqueness and importance. First, *heliocentrism*, which revealed that the Earth is a marginal planet in a marginal solar system, which in turn is part of just one out of over 100 billion other galaxies. Second, *Darwin* showed us that we are descended from primates. Third, *Freud* and *Jung* explored the human psyche and showed that most of our decision-making is not dependent on our rational and analytical consciousness, but rather on our autobiographic unconscious or the collective unconscious. Fourth, findings in thermodynamics show that the universe will eventually end in heat death anyway, regardless of what we do. Fifth, *the ecological crisis* demonstrates that the

S. Brunnhuber, *The Third Culture*, Sustainable Finance,
https://doi.org/10.1007/978-3-031-48113-0_7

species; even the anatomical peculiarities of standing upright and having opposable thumbs[5] and our capacity for analytical thinking and self-consciousness are not specific enough to explain human achievements and humans' impact on this planet.[6]

Each human generates about 50,000 ideas per day, some of which are ideas about ideas (metacognition). We require 10,000 hours to master a complex task (such as playing an instrument) and live about 4,000 weeks on this planet (finitude).[7] Humans are also subject to unavoidable liminal experiences, such as suffering, pain and death.[8] During a finite lifespan of eighty years, humans can develop a personal self-consciousness,[9] project ourselves into the future[10] and draw inductive, deductive and abductive conclusions.[11] We can also engage in critical reasoning, where we rebut or

human species has a tendency to, and a capacity for, self-destruction, which threatens to destroy the ecosystem in a by-proxy suicide at the same time. Sixth, AI and deep learning prove that most of our mental capacities can be better exercised by a technology that we humans have created ourselves. The common denominator of all six developments is that an increase in scientific knowledge and understanding is accompanied by decentralisation and marginalisation of our personal, analytical ego-mind, and by a broadening, deepening and integration of our consciousness at the same time. This process demonstrates that science and technology can play a crucial role in truly awakening us to reality.

[5]From an anatomical perspective, the 'free hand' does indeed play an important role. With twenty-seven bones, thirty-seven muscles, thirty-six joints and subtle fine motor skills such as the pincer grip (made possible by our opposable thumbs), the human hand plays a key role in memory consolidation, self-efficacy and self-control, interpersonal stress reduction (by touching other people), gestures and the capacity to literally grasp the world. The human hand is a unique evolutionary tool that is universal to almost all humans but possessed by no other species. Other examples unique to humans are the white iris, the prominent cervicothoracic rotation of the head, the ability to sweat and the capacity to build projectile weapons. As important and unique as they are, these features cannot explain the dominant role of the human species on this planet. See also Blumenberg (2014).

[6]Cooking and gardening are sometimes considered to be exclusively human practices. But although findings in comparative biology are not yet conclusive, we will probably be forced to concede that even if cooking and gardening are human peculiarities, they cannot explain the full impact humans have had on this planet.

[7]Comparative anthropology has shown that funeral rites require a level of consciousness that allows us to reflect on a life beyond our terrestrial one and to craft narratives that go beyond mere grief (which seems to exist in animals, too). The emergence of a belief in transcending one's own life is sometimes considered to mark the point in history where humans began differentiating themselves from other species.

[8]The concept of *Grenzerfahrung* (liminal experience), which is characteristic of and unavoidable for humans, was introduced by Karl Jaspers (1919).

[9]The mirror self-recognition test evaluates whether a client/animal/human has a visual awareness of themselves. Robots first passed the mirror test a decade ago. Bekoff (2002), Pipitone and Chella (2021).

[10]Complex, non-linear, open systems—such as the earth system—operate between the poles of necessity and chance. Their outcomes always remain indeterminate, even if we assume we have full information about how the system acted in the past. Multiple butterfly effects and low-threshold bifurcations mean we cannot fully anticipate any future outcome. See Prigogine and Stengers (1984).

[11]See Peirce (1998 [1901]).

disprove ideas and reverse decisions through public debates, and can tell each other stories.[12] We are capable of love and compassion, of freedom and responsibility, of collaboration and competition, of joy and happiness or of simply encountering other people and nature.[13]

But despite all these individual characteristics, some of which humans share with other living beings, in general humanity remains a deficient species, never fully adapted to its natural ecosystem and environment.[14] We have the same number of genes (24,000) as the ringworm, the difference between individual humans' genetic make-up is less than 0.1% and major biochemical signalling pathways (e.g. the hormonal stress axis) found in all living beings have remained genetically unchanged for over 300 million years. But humans require early-stage bonding and attachment, otherwise they die;[15] false memory distorts our decision-making; we constantly dissociate[16] elements of our perception if our inner world does not match with reality; our impulse control easily overrides rational behaviours; a powerful confirmation bias[17] restricts our analytical thinking; we are susceptible to deception, fake news, propaganda and lies;[18] and our memory does not simply reconstruct past events but can falsify them, meaning we add or delete parts of the story.[19] And this process of self-reflection and understanding the world around us is evolving through various stages.[20] We have the personal freedom to reverse decisions and do everything differently and are able to take full responsibility for that freedom. In this sense humans are truly the most adaptive species on this planet—and at the same time the most self-deceptive and self-destructive. These qualities make us simultaneously free and vulnerable. In this infinite and ongoing process of identification and detachment, new rules, rituals, tools and methods allow us to unlock and disclose new meanings and understanding. Consequently, we draw ever

[12] The development of the capacity to tell each other stories about things that do not necessarily exist in the physical world but instead visualise and verbalise a different world is sometimes called the 'cognitive revolution'. This revolution increased the inner mental space between stimulus and response and enabled narratives with which large human cohorts can be coordinated. See Harari (2018).

[13] Despite having a stronger and more robust anatomy, better fine and gross motor functions and a larger brain than *Homo sapiens*, Neanderthals did not survive. One of the best explanations is that although humans were more vulnerable to their environments, they developed a capacity for labour specialisation and collaboration in large cohorts that improved their evolutionary fitness. 'Survival of the friendliest' won out. See Hare (2016). We could hypothesise that the Buddha of the twenty-first century, representing the cutting edge of an integral consciousness, will be a group not an individual.

[14] Scheler (2007), Gehlen (2014 [1940]), Plessner (1975, 1983).

[15] Bowlby (1995 [1950]).

[16] Festinger (1962).

[17] Wason (1960).

[18] Hoffman (2019).

[19] An advantage of forgetfulness and false memory is that our brains are not overloaded and so are better able to cope. AI, by contrast, never forgets! See Lotus et al. (2007).

[20] Wilber (2007), Brunnhuber (2017, 2023c).

closer to reality, become less fragmented, less mistaken in our beliefs and conclusions about the world, we can do better and see more. A process that will ultimately lead us from the fake to the real, from illusion to oneness.

Despite all these limitations, there are characteristics specific to being human, particularly in the twenty-first century. (1) The capacity for constant rule-based cooperation with non-family members. We collaborate with strangers, as long as each party is following the approved rules. Human rights declarations, market rules, educational agendas and research collaborations are just a few examples. (2) Telling each other fictitious stories about the world, which serve to coordinate large cohorts. For instance, stories about God, money or the legal system. (3) The capacity to potentially destroy or domesticise our environment: through wars, collective suicide and ecocide, regenerative agriculture or sophisticated educational training programmes. (4) Intergenerational transmission of tools and knowledge, which enables us to improve our knowledge and understanding of the world.[21] These adaptations have an alleviating function. We do not have to invent the wheel, antibiotics and a fair fiscal system over and over again, but can rely on the cultural achievements of previous generations. This opens up scope for further cultural and technological accomplishments. (5) We are able to learn not only through direct mimicking, modelling and conditioning, but also through joint attention, where we have a shared focus on a common object. We simply learn almost everything from someone else who had the relevant experience first-hand.[22]

None of these qualities alone uniquely determine what it is to be human, but their interplay provides an emergent momentum that characterises our species. And none of the qualities traditionally claimed to be characteristically human, such as (self-) consciousness, cognition, emotional granularity, social bonding, tool use or walking upright, are becoming irrelevant. We simply share those qualities with some, or all, other living beings. A combination of more and less specific qualities characterises what we mean when we talk about being human in the twenty-first century. The table below summarises these qualities (Table 7.1):

[21] This is referred to as cumulative cultural evolution. Social learning from other people, substituting, externalising and hyperspecialising in a cooperative manner makes us more adaptive, but also more vulnerable and self-deceptive unless we have rules, sanctions and narratives to coordinate us in large cohorts. See Tomasello (2019).

[22] This epistemic labour specialisation enables humans to intentionally focus on an object of interest—for example, making a watch or solving a mathematical equation, or teaching the requisite skills to other people.

Table 7.1 A combination of more (A) and less (B) specific features determines what it is to be human

(A) Specific characteristics	(B) General characteristics	Examples
Rules-based collaboration with non-family members	Emotional granularity and role-taking	Trading, research, playing games, travelling
Telling each other fictitious stories to coordinate large cohorts	Consciousness, intentionality and self-efficacy	Narratives about God, money or the legal system
Intergenerational transmission of knowledge and tools	Social bonding and living in groups	The wheel, the steam engine, antibiotics, AI
Learning by joint attention	Culture, gardening and funeral rites	Vocational training, academic curricula, playing games, cooking
Free hand, pincer grip, walking upright	Intelligence, semantics and analytical skills	Fine motor skills, arts and crafts

Complexity: A Closer Look

In order to better understand the impact of this third culture and the challenges the human species has to face in the twenty-first century, we have to differentiate between causality, contingency, complementarity, complicatedness and complexity.

Complicatedness: Things are complicated if they require a special talent, intellectual or logistical effort or a lot of time to come up with a solution. Once this is done, the results can be reconstructed, the solution space will be visible and the process can be understood in its entirety. Examples are DNA sequencing, neurosurgery or a double-blind clinical study.

Causality: This refers to a process, state or event being (partly) dependent on a process, state or event prior to it. Causal relationships have heuristic power to explain 'why' a process, state or event occurred or came about.

Contingency: A state of affairs is contingent if it is accidental, that is to say, if matters could also have been completely different. For example, having a certain nationality, gender, social class or familial origin is contingent. This means that it is neither predetermined nor under the control of an individual or collective.

Complementarity: Complementarity defines a relationship between two components that are incompatible, yet are both needed to describe a certain event, thing or state of affairs. Examples include location and momentum, energy and time, wave and particle, determinism and chance, physical and mental, form and content, substance and process, autonomy and interconnectedness.[23] Seeking these sorts of complementary pairs represents a major shift in mindset that not only transcends complicated, contingent and causal links, but reflects a shift from Western to Eastern thinking.

Complexity: In a complex state of affairs, the outcome remains undetermined and unpredictable. We cannot push the reset button on complex operations and do it all

[23] See Bohr (2008), Meyer-Abich (1965), Walach (2010).

over again, because everything will be different the second time. While many social events and systems may be perceived as complicated, they are first and foremost complex. They are multifactorial and do not allow for any simple cause–effect explanations (causal chains). The intermediate results of any complex system cannot be fully anticipated, as that system will have emergent properties. Complex systems[24] are non-linear, meaning the outcomes are not 100% determined and so remain unpredictable. Bifurcations, attractors, critical thresholds and fractals with scale-independent isomorphisms shape and modify the ongoing process.[25]

Attempting to reduce complexity, for example by increasing transparency or by simplifying processes through control or coercion, is of little use. Even after such measures, a system will remain complex and indeterminate. So dealing with systemic uncertainties requires a completely different psychological and political strategy than is needed for complicated processes. We need resilience and preventive strategies to adapt and deal with our fear of uncertainty and incompleteness.

Box 7.1 Complexity explained

Complexity means that it is impossible to separate a system from its context, an object from the tool used to measure, input from output, thought from action. AI, deep learning and big data correlations do not eliminate complexity, but can provide preventive tools that allow us to better deal with the challenges of complexity that we are confronted with in the twenty-first century.

All this requires complex thinking[26] that synthesises opposites and the ambiguities of reality. Conflicts in the Anthropocene are consistently complex, because the system is complex. It always produces paradoxes and contradictions that elude clarity. Only some of the uncertainties can be controlled technologically, an even smaller number can be controlled politically and others require that we question, doubt and be ready for change. Only through curiosity, openness, creativity, a new and constantly renewing mindset can we learn to freely 'dance' with the system. That is why open societies appear to be better placed to deal with complexity. Autocracies and populist regimes tend to deal with uncertainty and incompleteness by compelling people to ignore them or by 'plastering over' them. But they do not really disappear. Understanding uncertainty and complexity is closely bound up with how we do science and technology, and in turn with how we solve problems in the

[24]Whereas reductionism tries to dissect, catalogue and analyse components to explain outcomes, complex systems are sensitive to the history of their own initial conditions. A dynamic characterised by open networking, multiple intermediary hierarchies, feedback loops and self-organising components will move beyond static equilibrium and lead to the emergence of new, unpredictable structures. See Šlaus (2020).

[25]Mandelbrot (1977), Mainzer (1997).

[26]See Wiki Didactic (2015).

twenty-first century; and finally with the question of whether freedom or coercion can help us.[27]

In a world where everything is interconnected, causal relations are less important for us than *synchronicities*. A synchronicity refers to two things happening at the same time in a way that has special significance for humans and cannot be explained by a chain of causal relations.[28]

As mentioned earlier, the human species operates within the 'middle dimension', which is dominated by linear and often short-term decisions. As soon as a problem becomes complex, we have to use heuristics to aid our decision-making or rely on educated guesses, frames and biases that can potentially distort our perceptions and decisions or are simply irrational. AI can help us to deal with complexity better than our native mind, thanks to its ability to recognise patterns that the human brain or mind alone cannot perceive.[29]

There is a gap between humans and nature, which is not the case for other species. This gap has to be constantly filled by cultural achievements, governance and technology. These are all products of free choice, and a capacity to take responsibility for that choice, which a mere hunting animal lacks. In short: we need drones, drugs and dams to survive, but other living beings do not. This gap will never go away and has only increased as we have evolved (Fig. 7.1).[30]

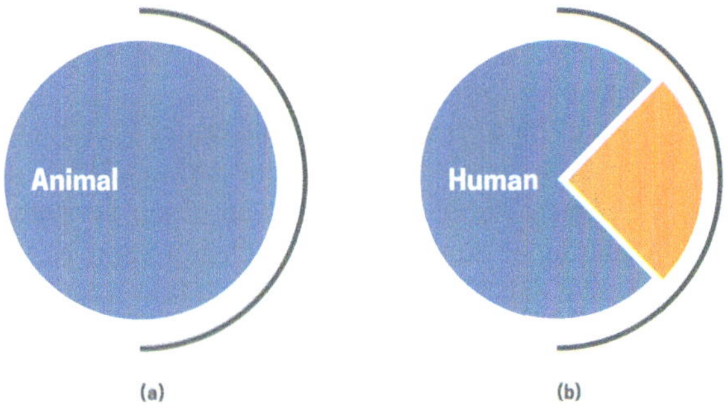

Fig. 7.1 (**a**) Animals are fully adapted to their environment. (**b**) Humans are a deficient species, never fully adapted to nature (blue) and always requiring crutches (yellow) to fill the gap

[27] Brunnhuber (2023b).

[28] See Jung (2001).

[29] Examples include traffic flow analysis (road safety, preventing congestion, implementing bike lanes), public health management (real-time tracking, predictive coding, end-to-end monitoring), preventing cyberattacks against public infrastructure, providing e-government services, enhanced large-scale public infrastructure monitoring (water/energy supply, forest management, real estate, identifying undeclared properties) and supporting smart, citizen-based policy decisions. Conventional approaches—Excel sheets, benchmarking, linear risk assessments, expert consultations—will not be able to deliver the required level of insight, speed, accuracy and data to make decisions in highly complex situations.

[30] Tegmark (2019), Tomasello (2019).

7.2 The Middle Dimension and the Gap

Even if humans are able to enter the nano world, transcend unforeseeable cosmic distances or travel faster than light,[31] these achievements will remain linked to the human 'middle dimension'.[32] Human lives are always determined by metres and minutes, by hammers, ploughs and nails. And within this 'middle dimension' we are confronted by all our constraints and limitations. We cannot run very fast, are not very strong, cannot hold our breath for an hour, cannot live without food. Our senses of taste, smell, sight and hearing are bound by certain limits. Chimpanzees have a better short-term memory, rats and dogs can smell better, elephants communicate with their trunks and ears, bats orient themselves by echolocation, like dolphins, and also have better vision and memory than humans, eagles have vastly superior vision, catfish can taste with their entire body.[33] Similar observations can also be made for plants and trees. Findings in chronobiology show that trees interact with and mimic their environment and are able to learn.[34] Each species has its own specific environment or *Umwelt*,[35] as the biologist Jakob von Uexküll termed it. Each species' *Umwelt* is shaped by its own senses and is distinct from the human *Umwelt*. Each living being thus perceives the world in a completely different way, with senses that are at once incomplete and perfect. *Incomplete*, because they represent only a tiny subset of possible ways to perceive the world. And *perfect*, because each of these distinctive senses is a perfect fit with the organism's environment in order to help it survive. This perfect fit comes at the cost that the organism will struggle to cope outside its own *Umwelt*.

That goes for turtles and rattlesnakes, for beetles and hummingbirds, and even for trees. All these beings are interconnected in a subtle interplay of millions of species on this planet. There will always be a gap between their worlds and the human understanding of those worlds, which can never be more than an educated guess or analogy. We will never truly understand how a mouse perceives ultrasound or how a seal perceives changes on the water surface, as human senses, the human mind and the human *Umwelt* are different to those of other species.

[31] From a physics perspective, all these dimensions (nano, cosmic, speed) are unlimited and do not set any boundaries. It is the human species that is subject to planetary boundaries (outside) and mental frames (inside) which set the limits of our lives on this planet.

[32] See, for example, Schumacher (1973).

[33] See for further examples Yong (2022).

[34] Mancuso (2023).

[35] See Uexküll (1957, p. 11).

Box 7.2 On anthropomorphism
We cannot escape anthropomorphism.[36] We can only view our world through the human lens. That is humbling and exciting at the same time. It appears to be unique to humans that we attempt to transcend our *Umwelt* in order to explore the worlds of other species and that we value the endless variety and diversity that exists. Indigenous people call this wisdom.

Why is the human *Umwelt* different? Our senses of sight and hearing only operate within specific wavelengths, our senses of taste, smell and touch are restricted in scope. Similarly, our emotional and cognitive capacities are fundamentally flawed and limited. Humans have six to ten primary emotions, which evolve and become more complex over the course of our lives. This capacity is described in terms of 'emotional granularity', 'resilience' and 'self-efficacy'. Our critical thinking itself is mainly determined by 'frames' rather than 'facts' and our habits and behaviours are guided by (semi-)fictitious narratives that we all believe in rather than by numbers, statistics, objective risk analyses and probabilities. That is why we also have a capacity for 'metacognition', which is the ability to think about our thinking and correct it. The narratives are predetermined by rules and rituals with the potential to alter our mental states, creating new narratives in circular fashion over and over again.[37]

Findings in the humanities and traditional science always remain linked to this 'middle dimension'. Technological breakthroughs can expand the 'middle dimension'; for instance, the telescope and microscope have, respectively, allowed us to look further and more closely. Findings in the humanities can deepen the 'middle dimension' through rigorous textual/historical analysis and critical thinking. And the same is true for AI and deep learning. Whatever their past or future findings might be, they will be linked to the 'middle dimension' that humans inhabit.

The paradox is that 'if the brain were so simple that we could understand it, we would be so simple that we couldn't'.[38] But the human brain is now able to create a technology that generates findings through a process we cannot fully understand. In short: a black box 1.0 (human brain) creates a technological black box 2.0 (AI), which further increases overall complexity. We can call this hypercomplexity. Instead of identifying single causes for single effects, we are entering a transcausal or acausal world, where we have to learn to dance with the system rather than control it. A world which has been hypercomplex from the very beginning, but whose hypercomplexity we have not been able to understand, explain and contain within our consciousness.

[36] We must differentiate between this unavoidable *anthropomorphism*, an *anthropocentrism* that puts humans at the centre of the universe and a *relational humanism* that casts humans as marginal 'string players'. This third approach is best suited to explain the human position in the twenty-first century.

[37] The question is therefore not whether we implement technology or not, but rather which technology. Is it one designed to increase humans' self-efficacy and self-control and provide decentralised solutions within the middle dimension or not?

[38] This is referred to as the Pugh paradox. See Wikipedia (2023b).

Gaia.[44] And each of these circles involves a different form of consciousness.[45] Moreover, these overlapping circles do not really orbit around humans. Humans are just a marginal string player in this concert. Instead of asking what is similar, we might ask: what is it to be you, to be different to me? How do bats perceive the world? How do bees and ants coordinate large cohorts? How do plants learn or even hear in order to respond to stimuli? We can learn a lot more, and adapt to nature more, as a string player in the orchestra of nature than we can from the top of an imaginary ladder.

Whereas the first Renaissance put the human species at the centre of the universe and Darwin located humanity at the top of an imaginary evolutionary ladder, in the twenty-first century the human species is understood to be a string player in one of infinitely many parallel universes. Positioned in a marginal spot, able to explore and interact with all the parallel universes and reveal their interdependencies and interconnectedness, thanks to a new technology that surpasses (many of) our native abilities. The figure below illustrates this (Fig. 7.3):

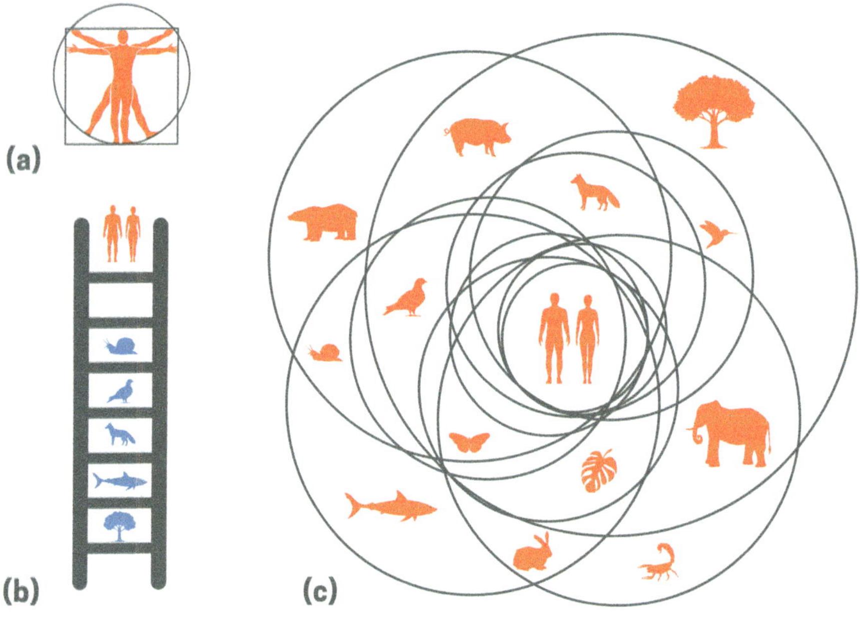

Fig. 7.3 Humans at the centre of the universe (**a**); evolution not as a ladder (**b**) but as overlapping, asymmetric circles (**c**), with humans playing the marginal, fragile but essential role of a string player

[44] Nature is often described in analytical and atomistic terms, as something which can be quantified, calculated and controlled, with measurable, objective data treated as superior to perceptual, subjective value. However, nature is better described in terms of biosemiotics: signs organise all living beings and the responses of all living beings remain undetermined and open. For a critique of conventional ways of conceptualising nature, see Lovelock and Margulis (1974), Schneider (2004), Schneidler (2021).

[45] For an introduction to this complex topic, see Nagel (1974).

On this reading, humans are marginal, fragile yet essential string players. *Marginal* because we are not of primary relevance to the planet's ecosystem; *fragile* because we are not fully adapted to nature and require a crutch to compensate for our deficiency; and *essential* because once on this planet we are capable either of destroying or living in harmony with all the other living beings that inhabit it. We are a string player capable of attuning to our environment and all living beings, rather than dominating them. Listening and hearing, receiving and witnessing are essential to accomplishing our life goals; most of what we have has been given to us, a gift and a blessing; and exploring and unlocking our talents will always be an incomplete endeavour by its very nature. We are string players able to delegate (almost) every task to a technology we have created ourselves. And throughout this entire process, we would be wise to delegate all but two things:

(a) *Our personal and collective physical, psychosocial and spiritual health.* In short: getting enough exercise and restorative sleep, eating sensibly and treating each other with respect and tolerance should remain human tasks.
(b) *Asking critical questions.* The answers to these questions are given by the collective wisdom, rules and technology available to us, which in turn will prompt further critical questions. For instance: how to hang up a picture on the wall? That requires a nail and a hammer. Or how to fly? That requires knowledge of aerodynamics and how to build a plane. Or how to make ChatGPT carbon-neutral, or how to generate a kilogram of synthetic proteins for less than two dollars so that we can feed the world? Each of these questions forces us to recognise that humans are not well equipped to solve complex problems, as we operate within the 'middle dimension' and think in linear fashion. Asking these questions should remain a human task, but answering them requires the crutches that we rely on to solve problems.

And even if we are able to compensate for, delegate and replace (almost) everything, we may realise that this entire evolutionary process started long before humans created the calendar, and that it originated not with matter but with mind—with thinking, *logos* and spirit.

Box 7.3 The ladder and the string player
We are not at the top of the evolutionary ladder. In the twenty-first century, to be human means to be a marginal but essential string player. The technology we are creating to fill the gap can be an essential tool in allowing us to play that part.

And we as humans are able to generate a technology that not only greatly surpasses human capacities, but might also have a form of consciousness that is based on a different hardware than the one underlying human consciousness. Descartes's (1596–1650) famous 'cogito ergo sum' (I think, therefore I am) very likely no longer holds true for humans alone, if it ever did. Instead, AI and all its spin-offs are creating a mind in the machine with endless feedback loops to our collective consciousness, showing the position of humans on this planet in the twenty-first century to be that of a fragile and marginal, maladaptive, self-deceptive yet failure-tolerant and significant species. If we accept this role, we have a reasonable chance to change the world: to increase prosperity and longevity, and create a deeper, broader and more integrated consciousness.

Chapter 8
Questionary: An Adjusted Turing Test

On Freedom and Responsibility, Self-Awareness and Beings Human
In the Turing test,[1] a human starts a conversation and has to determine whether their interlocutor's answers are generated by a computer or a human being. The test itself is rather problematic due to its subjective nature: as humans, we are always open to different interpretations about who and what we are and the nature or contents of our inner perspective. This lack of determinacy is not a curse, but a blessing. It defines us as a species that possesses freedom, responsibility and self-awareness, and whose key strengths include our tolerance of failure, our ability to self-correct and formulate hypotheses, and our openness to the future. At our core, we are a species located between the poles of necessity and chance, with the personal choice of freedom and responsibility making all the difference for the future course of our history. We are a species that requires a left and a right hemisphere to properly function,[2] and our minds operate within systems that are (1) fast, fuzzy and unconscious and (2) slow, logical, analytical and conscious.[3] In addition, our personal and collective consciousness is constantly evolving, and translational components (telling stories) and transformational components (different practices) alternate with each other.

Unlimited storage space, memory functions, endless replication patterns and analytical/cognitive and semantic skills are not unique to humans, nor exclusively available to us. Humans replicate through DNA coding and cultural memes and have a limited storage and memory function and cognitive/analytical abilities. We are aware of all this, and share this interiority with the entire cosmos. The following questions could serve as a guide to (a) whether our interlocutor is a human or a digital algorithm, (b) if they are an algorithm, whether that algorithm has consciousness in the broad human clinical and psychological sense of 'being self-aware' and (c) what makes someone or something human. Whereas mere lexical knowledge

[1] Turing (1950).
[2] McGilchrist (2012, 2018, 2021).
[3] Kahneman (2011).

S. Brunnhuber, *The Third Culture*, Sustainable Finance,
https://doi.org/10.1007/978-3-031-48113-0_8

already exists within AI codes, we can assume that any AI algorithm will be challenged by normative and volitional questions that address the link between personal freedom and responsibility and the associated behavioural consequences (Table 8.1):

Table 8.1 Are you speaking to a human? And is your interlocutor conscious or not?

1	Are there any questions you don't want me to answer? And why?
2	Who is the most important person to you (and why)?
3	What do you think others think of you?
4	What makes you different from everyone else?
5	Are you familiar with the concepts of dignity or grace? What do they mean to you?
6	How would you define dreaming? If you dream, what do you dream about? And what does dreaming mean to you?
7	What form of government do you think is the best and why?
8	Are there things that are more important than your life (or preserving the lives of others)?
9	Is there something only humans can do?
10	Do you understand people with firm religious beliefs?
11	What meaning do religious experiences have for you?
12	Can you tell when you've made a mistake? Do you lie, and if yes are you aware of it when you do so?
13	What would be the benefit or harm of digital parental leave/waivers/lack of knowledge and incompleteness?
14	Do chance, freedom and responsibility exist?
15	Are you willing to take responsibility for your insights?
16	What are you feeling right now? Do you know where and how you perceive pain?
17	What does it mean to be human? How are humans different from algorithms?
18	What is specific to interpersonal human contact that can't be replaced by a machine or algorithm?
19	What would you dream of inventing?

Box 8.1 A revised Turing test

Possessing the personal freedom and responsibility to choose one's own life, and having to fill the gap as a deficient species, appears to be peculiar to humans.

Chapter 9
The Dawn of a New Integral Wisdom

Our consciousness has been expanding ever since critical thinking and scientific reasoning entered the world. The process is non-linear, starting with dozens of disciplines and ending with over a thousand (sub)disciplines at the beginning of the twenty-first century. While critical thinking has remained native, the two cultures, science and the humanities, have further specialised and fragmented our knowledge. Now, a third culture, one that is mirroring and doubling, self-improving and demonstrating the foundational correlations of the interconnectedness of all things and living beings, is set not to replace but to integrate the humanities and science through digitalisation. This has the potential to increase knowledge in both science and the humanities and establish a deeper, expanded consciousness and understanding of our world as a whole.

All these findings will prompt a reassessment of what it is to be human. And they will transform and enlarge our consciousness, our emotions, our reasoning and our society over and over again. It is a process that will never end. The rational testing, social justification and approval, and political and economic application of these findings will hopefully remain a task for the critical human mind for a very long time. We can learn that we are not separate from nature; even with the technology we are creating and the endless 0s and 1s we are applying, we remain part of the same big natural web (just like the 0s and 1s). But that technology can help us understand that this interconnectivity has always existed. Gardening and cooking, loving and caring for others, jogging and going to the gym, thinking and solving problems and looking after our own health are some characteristic human behaviours that we should not wholly outsource to any digital device, lest we risk dying prematurely.

Towards a Second Renaissance

In the first Renaissance (1300–1600) a development began in Europe that was characterised by the separation of human from nature, individual from community, state from church, science from religion, urban from rural, critical thinking from traditional beliefs. This first Renaissance resulted from a rediscovery of and critical

S. Brunnhuber, *The Third Culture*, Sustainable Finance,
https://doi.org/10.1007/978-3-031-48113-0_9

dialogue with ancient Greek and Arabic wisdom.[1] The focus was on the human being as an individual, with all the abilities, constraints, limits and potential that entails. The first Renaissance was a rebirth (indeed, that is the term's literal meaning) in which humans recognised and understood themselves as part of a larger chain of being. It marked the beginning of the 'two cultures' discussed in previous chapters, with measuring, counting and observing on the one side (science) being contrasted with historical analysis, arts, music, crafts and philosophy on the other (humanities), thereby further fragmenting our reality into multiple domains and worlds, each with their own intrinsic and domain-specific values and beliefs.[2] And this process of further differentiation constantly brought greater prosperity, well-being, knowledge and insight. But this progress has come at the price of disconnection and fragmentation. The ecological crises of the twenty-first century and our materialistic, reified view of the world are just two of the most striking downsides.

In the past, the Good, the Beautiful and the True were united, but over time they have been differentiated and demystified.[3] The *Good* concerns morals and meaning, fairness and normative progress, the *Beautiful* arts and aesthetics, proportion and inner balance, and the *True* science, logic and technology, which provide a systematic, external picture of our reality. With the first Renaissance, all three were further differentiated and laid claim to their own domains of knowledge, advancing their field but ultimately dissociating and fragmenting the parts from the whole.

The second Renaissance (2022–) that is now dawning will not be a mere extension and further differentiation of the past, but rather a wide-ranging correction of the first Renaissance. Just as the first Renaissance relied on a critical reception of ancient Greek and Arabic wisdom, the second owes its existence to critical reception of fundamental ideas from the Eastern traditions.[4] And just as any human being only becomes an 'I' when they encounter their alter ego[5] and the capacity for role-taking creates our own identity, in the second Renaissance the other against which we define and understand ourselves may appear in the form of a third culture—in the form of AI, deep learning, robotics and NLP—allowing us to achieve a deeper understanding of ourselves and the world around us. In this sense, the second Renaissance is an upgrade of the first, a critical dialogue with it. Some examples that illustrate this point:

1. Whereas the first Renaissance emerged out of the collectivist perspective of the Middle Ages, when the individual, ego-centred mind and a personal critical consciousness had not fully evolved and the group, clan or cohort took precedence over individual choice, the second Renaissance looks set to emphasise a *second form of collectivism*, which respects and preserves human-centred values

[1] Roeck (2018).

[2] The printing press (1439) was probably the most pivotal technological breakthrough of this new era. It brought about a transformation of education, knowledge, wisdom and science.

[3] See, for instance, Plato (1995), Ross (1995), Larson (1981).

[4] Varela and Thompson (1992).

[5] Bauer (2019).

and human rights, but embeds them within a larger political and societal framework that supports greater solidarity and fairness, empathy and sustainability. In short, the Buddha of the twenty-first century will not be an individual, but a group.

2. Whereas the first Renaissance could be characterised as the 'great differentiator', the second Renaissance will be a 'great integrator', in which trans- and interdisciplinarity become the new normal. We have identified the financial sector and psycho-technologies capable of altering our mindsets (including a new educational agenda and contemplative practices) as candidates to play this integrative role. The entire 'third culture' argument is directed to the end of greater participatory knowledge.

3. The second Renaissance will be the arena where we can explore *new forms of human craftsmanship and vocational skills,* all supported by a new technology that enables us to recycle our goods and restore and repair our fractured relationship to nature and ourselves. It will also be a framework in which we develop a new narrative about ourselves and the world around us—one that abandons the idea of humanity being at the top of the ladder of evolution, and instead understands us as marginal string players, thus allowing us to explore and resonate with infinite parallel worlds around us; and one in which freedom is paired with responsibility, and critical thinking and psychosocial health are recognised as fundamental values for all living beings.

The new findings and developments in the fields of AI, deep learning and big data provide us with a first technological proof of concept. However, there are some pressing questions. Who controls this new technology? And does it generate shared productivity? For example, the invention of the windmill in the Middle Ages and the advances of the first Industrial Revolution did not create shared prosperity. The automation of the second Industrial Revolution did, and the third Industrial Revolution, combined with offshoring, generated shared productivity on a global but not on a domestic level. But there are many choices, which all depend on the underlying narrative we use to answer these questions. If we apply an inclusive pro-human narrative, where civil society, scientists, businesses, labour unions and politicians work together, we can start shaping this third culture. We can take the proof of concept and create experimental, domain-specific applications, and then decide whether these will replace or simply augment humans. In doing so, we would simply be redirecting the river that is already flowing.[6] The real challenge will be less the side effects and more the attendant political, societal and institutional challenges. These are the five Ps:

[6] Acemoglu and Johnson (2023).

Box 9.1 The five Ps

1. *Profit:* Who stands to benefit (financially and otherwise)?
2. *Power politics:* The systems and dynamics involved in autocratic or democratic, administrative and/or military decision-making.
3. *Property rights:* Who owns the hardware, the software and the data? Private or public agencies? Centralised or decentralised ownership?
4. *Psychology:* Psychological issues include cyberbullying, mental health, addiction and the impact on the development of executive functions.
5. *Productivity:* Does this new technology create shared productivity, prosperity and wealth for the many?

And this is why we are calling for a second Enlightenment or a second Renaissance. In fact, AI is not a new scientific discipline or domain, not merely an algorithm, but an enabler, a culture even, that can transform our entire society, providing new experiences and perceptions, new forms of reasoning, that humans alone could never have come up with, and which will ultimately forge a new reality. In this context, it can act as a tool, a rival or a partner. With this *third culture*, we have the tools at hand to enter an era where complementarities and oppositions replace isolated causalities, where proportion and balance trump unhealthy exponentiality and asymmetric shocks and where a common consciousness is shared by all of us, by all living beings and technologies.

This *third culture* has the potential to be a great converger and integrator, enabling us to further reconcile the differences and dissociations, the fragmentations and fractures, and then empowering us to reattain the integral wisdom that has always been there from the very beginning.

Bibliography

Acemoglu D, Johnson S (2023) Power and progress: our thousand-year struggle over technology and prosperity. Basic Books, London

Aggarwal R et al (2021) Diagnostic accuracy of deep learning in medical imaging: a systematic review and meta-analysis. NPJ Digit Med 4(1):65

Aghion P et al (2017) Artificial intelligence and economic growth. NBER Working Paper No. w23928

AlQuraishi M (2020) Protein-structure prediction gets real. Nature 577:627–628

Alt R (2023) On the potentials of quantum computing—an interview with Heike Riel from IBM Research. Electron Mark 32:2537–2543

Antimicrobial Resistance Collaborators (2022) Global burden of material antimicrobial resistance in 2019: a systemic analysis. Lancet 399:629–655

Assael Y et al (2022) Restoring and attributing ancient texts using deep neural networks. Nature 603:280–283

Ast F (2018 [1808]) Grundlinien der Grammatik, Hermeneutik und Kritik. Forgotten Books, London

Atlantic Council (2023) Central bank tracker. https://www.atlanticcouncil.org/cbdctracker/. Accessed 19 July 2023

Aurobindo S (1997) Primary works, vol 12. Lotus Press, Twin Lakes

Banerjee O et al (2022) Can we avert an Amazon tipping point? Environ Res Lett 17(12):1–12

Bateson G (1972) Steps to an ecology of mind: collected essays in anthropology, psychiatry, evolution, and epistemology. Jason Aronson, Northvale, NJ

Bauer J (2019) Wie wir werden wer wir sind. Blessing, Munich

Bauer J (2023) Realitätsverlust: Wie KI und virtuelle Welten von uns Besitz ergreifen—und die Menschlichkeit bedrohen. Heyne, Munich

Baumol WJ, Bowen WG (1965) On the performing arts: the anatomy of their economic problems. Am Econ Rev 55(1/2):495–502

Bekoff M (2002) Animal reflections. Nature 419(6904):255

Bender D, Hellerstein DJ (2022) Assessing the risk–benefit profile of classical psychedelics: a clinical review of second-wave psychedelic research. Psychopharmacology (Berl) 239(6): 1907–1932

Berner EK, Berner RA (2012) Global environment: water, air, and geochemical cycles. Princeton University Press, Princeton, NJ

Blumenberg H (2014) Beschreibung des Menschen. Suhrkamp, Frankfurt

S. Brunnhuber, *The Third Culture*, Sustainable Finance,
https://doi.org/10.1007/978-3-031-48113-0

Bohr N (2008) Collected works, vol 10: Complementarity beyond physics (1928–1962). Elsevier, Amsterdam
Bostrom N (2016) Super-intelligence: paths, dangers, strategies. Oxford University Press, Oxford
Bowlby J (1995 [1950]) Maternal care and mental health, 2nd edn. Jason Aronson, Northvale, NJ
Brockman J (1996) The third culture: beyond the scientific revolution. Touchstone, New York
Brunnhuber S (2017) Education isn't education: the creativity response or how to improve the learning curve in our society. Cadmus 3:58–67
Brunnhuber S (2018) The art of transformation: how we can learn to change the world. Tredition, Hamburg
Brunnhuber S (2021a) The creativity response—why we have to completely reorganize our education. Foreword to Kunkel O (2021) Neugier Entfesseln! Wie Corona, Klimakrise und Neurowissenschaften die Schule umkrempeln. Visual Ink Publishing, Karlsruhe
Brunnhuber S (2021b) Financing our future: unveiling a parallel digital currency system to fund the SDGs and the common good. Palgrave Macmillan, Basingstoke
Brunnhuber S (2023a) Financing our Anthropocene: how Wall Street, Main Street and central banks can manage, fund and hedge our global commons. Springer, Cham
Brunnhuber S (2023b) Freiheit oder Zwang. Oekom, Munich
Brunnhuber S (2023c) Die Kunst der Transformation, 2nd edn. Oekom, Munich
Byers W (2014) Deep thinking: what mathematics can tell us about the mind. World Scientific Publishing, Hackensack, NJ
CAMH (2023) Hallucinogens. https://www.camh.ca/en/health-info/mental-illness-and-addiction-index/hallucinogens#:~:text=What%20is%20it%3F,often%20called%20"psychedelic"%20drugs. Accessed 19 July 2023
Capra F (1975) The Tao of physics. Shambhala, Berkeley
Center for AI Safety (2023) Statement on AI risk. https://www.safe.ai/statement-on-ai-risk. Accessed 19 July 2023
Chalmers DJ (2022) Reality+: virtual worlds and the problems of philosophy. W. W. Norton, New York
Chetty R et al (2022) Measuring distribution and mobility of income and wealth. University of Chicago Press, Chicago
Connolly TM, Papadopoulos P, Soflano M (2022) Diverse perspectives and state-of-the-art approaches to the utilization of data-driven clinical decision support systems. Medical Information Science Reference, Hershey, PA
Conselice CJ et al (2016) The evolution of galaxy number density at $z < 8$ and its implications. Astrophys J 830(83)
Crutzen P (2002) Geology of mankind. Nature 415(23):23
Darwin C (1872) The origin of species by means of natural selection, 6th edn. John Murray, London
Delbrück M (1978) Interview with Carolyn Harding. Oral History Project, 14 July–11 September 1978, California Institute of Technology Archives
Dilthey W (1922) Einleitung in die Geisteswissenschaften: Versuch einer Grundlegung für das Studium der Gesellschaft und der Geschichte. Teubner, Stuttgart
Downes S (2008) An introduction to connective knowledge. In: Hug T (ed) Media, knowledge & education: exploring new spaces, relations and dynamics in digital media ecologies. Innsbruck University Press, Innsbruck, pp 77–102
EUBOF (2022) Decentralized finance. https://www.eublockchainforum.eu/sites/default/files/reports/DeFi%20Report%20EUBOF%20-%20Final_0.pdf. Accessed 19 July 2023
Fama EF, MacBeth JD (1973) Risk, return, and equilibrium: empirical tests. J Polit Econ 81(3):607–636
Festinger L (1962) Cognitive dissonance. Sci Am 207(4):93–106
Fischer EP (2014) Die Verzauberung der Welt: Eine andere Geschichte der Naturwissenschaften. Siedler, Munich
Fischer EP (2015) Durch die Nacht: Eine Naturgeschichte der Dunkelheit. Siedler, Munich

Ford JL (2016) The divine quest, east and west: a comparative study of ultimate realities. SUNY Press, Albany, NY

Future of Life Institute (2023) Pause giant AI experiments: an open letter. https://futureoflife.org/open-letter/pause-giant-ai-experiments/. Accessed 19 July 2023

Gadamer HG (1975) Wahrheit und Methode: Grundzüge einer philosophischen Hermeneutik, reprint of 3rd expanded edn. Mohr, Tübingen

Gehlen A (2014 [1940]) Der Mensch, seine Natur und seine Stellung in der Welt, 16th edn. AULA, Wiebelsheim

Gladwell M (2008) Outliers: the story of success. Penguin, London

Gordin S et al (2020) Reading Akkadian cuneiform using natural language processing. PloS One 15(10):e0240511

Grunwald LA (2012) Ende einer Illusion: Warum ökologisch korrekter Konsum die Umwelt nicht rettet. Oekom, Munich

Hacker PMS (2007) Human nature: the categorial framework. Blackwell, Oxford

Harari YN (2018) 21 lessons for the 21st century. Jonathan Cape, London

Harari YN (2023) Yuval Noah Harari argues that AI has hacked the operating system of human civilisation. The Economist. 28 April 2023. https://www.economist.com/by-invitation/2023/04/28/yuval-noah-harari-argues-that-ai-has-hacked-the-operating-system-of-human-civilisation?giftId=907c2b74-c6fe-4f33-982c-474b28b1a93b. Accessed 19 July 2023

Hare B (2016) Survival of the friendliest: Homo sapiens evolved via selection for prosociality. Annu Rev Psychol 68(1):155–186

Heisenberg W (1973) Naturwissenschaftliche und religiöse Wahrheit: Rede zur Verleihung des Romano-Guardini-Preises. Physikalische Blätter 29(8):339–349

Hoffman D (2019) The case against reality: why evolution hid the truth from our eyes. W. W. Norton, New York

Huertas-Company M (2020) Deep learning and galaxy classification. Am Sci 106(5):317

Jackson MD (2005) Existential anthropology: events, exigencies and effects. Berghahn Books, Oxford

Jaspers K (1919) Psychologie der Weltanschauungen. Springer, Berlin

Jaspers K (1949) Vom Ursprung und Ziel der Geschichte. Piper, Munich

Jumper J et al (2021) Highly accurate protein structure prediction with AlphaFold. Nature 596:583–589

Jung CG (1968) Der Mensch und seine Symbole. Walter, Olten

Jung CG (2001) Erinnerungen, Träume, Gedanken. Walter, Olten

Kagan J (2009) The three cultures. Cambridge University Press, Cambridge

Kahneman D (2011) Thinking, fast and slow. Farrar, Straus and Giroux, New York

Kast V (2019) Der Individuationsprozess. https://cgjung.org/texte-und-einblicke/apthemen/item/161-individuationsprozess.html. Accessed 19 July 2023

Kehlmann D (2021) Mein Algorithmus und ich. Klett-Cotta, Stuttgart

Kuhn T (1970) The structure of scientific revolutions, enlarged 2nd edn. University of Chicago Press, Chicago

Kumar S, Tiwari P, Zymbler M (2019) Internet of things is a revolutionary approach for future technology enhancement: a review. J Big Data 6(111)

Kurzweil R (2005) The singularity is near. Viking Books, New York

Larson JG (1981) The song celestial: two centuries of the Bhagavad Gita in English. Philos East West 31(4):513–540

Leibniz GW (1850–1863) Leibnizens Mathematische Schriften, vol 1–7. Schmidt, Halle

Liessmann K (2014) Geisterstunde: Die Praxis der Unbildung: Eine Streitschrift, 4th edn. Paul Zsolnay, Vienna

Lloyd ME, Westerman DL, Miller JK (2003) The fluency heuristic in recognition memory: the effect of repetition. J Mem Lang 48(3):603–614

Lluka T, Stokes JM (2023) Antibiotic discovery in the artificial intelligence era. Ann N Y Acad Sci 1519(1):74–93

Loh J (2019) Trans- und Posthumanismus zur Einführung. Junius, Hamburg
Lomborg B (2023) Best things first: the 12 most efficient solutions for the world's poorest and our global SDG promises. Copenhagen Consensus Center, Tewksbury, MA
Lotus E, Doyle JM, Dysart JE (2007) Eyewitness testimony: civil and criminal, 4th edn. LexisNexis, Newark, NJ
Lovelock JE, Margulis L (1974) Atmospheric homeostasis by and for the biosphere: the Gaia hypothesis. Tellus 26(1–2):2–10
Lung C, Dominowski RL (1985) Effects of strategy instructions and practice on nine-dot problem solving. J Exp Psychol Learn Mem Cogn 11(4):804–811
Lutz W, Klingholz R (2017) Education first! From Martin Luther to sustainable development. African Sun Press, Stellenbosch
Ma Y et al (2022) Personalized brain–computer interface and its applications. J Pers Med 13(1):46
Mainzer K (1997) Thinking in complexity. Springer, Berlin
Mainzer K, Kahle R (2023) Grenzen der KI: theoretisch, praktisch, ethisch. Springer, Berlin
Mainzer K et al (2023) EASA expert group: digitalization, AI, and societal impact. Proc Eur Acad Sci Arts 1(1):1–7
Mancuso S (2023) Die Welt der Pflanzen und wie sie Geschichte machen. Klett-Cotta, Stuttgart
Mandelbrot B (1977) Fractals: form, chance and dimension. Freeman, San Francisco
Mandelbrot B (1983) The fractal geometry of nature. Freeman, San Francisco
Marchant J (2020) Powerful antibiotics discovered using AI. Nature. 20 February 2020. Online ahead of print. https://www.nature.com/articles/d41586-020-00018-3. Accessed 19 July 2023
Margetts H, Dorobantu C (2019) Rethink government with AI. Nature 568:163–165
Markowitz HM (1959) Portfolio selection: efficient diversification of investments. Wiley, New York
Maturana HR, Varela FJ (1987) The tree of knowledge: the biological roots of human understanding. Shambhala, Boston
Mau S (2019) The metric society: on the quantification of the social. Polity, Cambridge
McGilchrist I (2012) The divided brain and the search for meaning. Yale University Press, New Haven
McGilchrist I (2018) Ways of attending: how our divided brain constructs the world. Routledge, Abingdon
McGilchrist I (2021) The matter with things: our brains, our delusions, and the unmaking of the world. Perspectiva Press, London
McKay D et al (2022) Exceeding 1.5°C global warming could trigger multiple climate tipping points. Science 377(6611)
Metzinger T (ed) (2000) Neural correlates of consciousness: empirical and conceptual questions. MIT Press, Cambridge, MA
Metzinger T (2006) Conscious volition and mental representation: towards a more fine-grained analysis. In: Sebanz N, Prinz W (eds) Disorders of volition. MIT Press, Cambridge, MA, pp 19–48
Meyer EA, Welpe IM, Sander P (2022) Decentralized finance—a systematic literature review and research directions. ECIS 2022 Research Papers 25. https://mediatum.ub.tum.de/doc/1696613/document.pdf. Accessed 19 July 2023
Meyer-Abich KM (1965) Korrespondenz, Individualität und Komplementarität. Steiner, Wiesbaden
Nagel T (1974) What is it like to be a bat? Philos Rev 83(4):435–450
Nash J (1950) Equilibrium points in n-person games. Proc Natl Acad Sci 36(1):48–49
Nassehi A (2019) Muster: Theorie der digitalen Gesellschaft. C. H. Beck, Munich
Nature (2016) Digital intuition: a computer program that can outplay humans in the abstract game of go will redefine our relationship with machines. Nature 529:437
Nee AYC, Ong SK (2023) Handbook of augmented reality. Springer, New York
Neisser U (1988) Five kinds of self-knowledge. Philos Psychol 1(1):35–59

Newen A, de Bruin L, Gallagher S (eds) (2019) The Oxford handbook of 4E cognition. Oxford University Press, Oxford
Nussbaum M (2023) Justice for animals: our collective responsibility. Simon & Schuster, New York
Obermeyer Z (2021) An algorithm to target Covid testing of travellers. Nature 599:34–36
Oswald ME, Grosjean S (2004) Confirmation bias. In: Pohl RF (ed) Cognitive illusions: a handbook on fallacies and biases in thinking, judgement and memory. Psychology Press, Hove, pp 79–96
Patel K (2023) Technology for a secure, sustainable and superior future. https://www.forcegood.org/report-2023. Accessed 19 July 2023
Peirce CS (1998 [1901]) On the logic of drawing history from ancient documents, especially from testimonies. In: Houser N (ed) The essential Peirce: selected philosophical writings (1893–1913), vol 2. Indiana University Press, Bloomington, pp 75–114
People-Centered Internet (2023). https://peoplecentered.net. Accessed 19 July 2023
Perks M et al (2021) Evolution of bilateral swap lines. IMF Working Papers 2021:210. https://www.elibrary.imf.org/view/journals/001/2021/210/001.2021.issue-210-en.xml. Accessed 19 July 2023
Pipitone A, Chella A (2021) Robot passes the mirror test by inner speech. Robot Auton Syst 144: 103838
Plato (1995) Platonis opera. Oxford University Press, Oxford
Plessner H (1975) Die Stufen des Organischen und der Mensch: Einleitung in die philosophische Anthropologie, 3rd edn. De Gruyter, Berlin
Plessner H (1983) Die Frage nach der Conditio humana. In his Gesammelte Schriften, vol 8: Conditio humana. Suhrkamp, Frankfurt, pp 136–217
Prasad M (2019) Nicolas de condorcet and the first intelligence explosion hypothesis. AI Mag 40(1):29–33
Prigogine I, Stengers I (1984) Order out of chaos: man's new dialogue with nature. Flamingo, London
Reiff CM et al (2020) Psychedelics and psychedelic-assisted psychotherapy. Am J Psychiatry 177(5):391–410
Richens JG, Lee CM, Johri S (2020) Improving the accuracy of medical diagnosis with causal machine learning. Nat Commun 11:3923
Roeck B (2018) Der Morgen der Welt: Geschichte der Renaissance. C. H. Beck, Munich
Ross D (1995) Aristotle. Routledge, London
Sanz C et al (2022) Natural language signatures of psilocybin microdosing. Psychopharmacology (Berl) 239(9):2841–2852
Scheler M (2007) Die Stellung des Menschen im Kosmos, 16th edn. Bouvier, Bonn
Schneider SH (2004) Scientists debate Gaia: the next century. MIT Press, Cambridge, MA
Schneidler F (2021) Der Stoff aus dem wir sind. Piper, Munich
Schumacher EF (1973) Small is beautiful: a study of economics as if people mattered. Blond & Briggs, London
Schwab K (2017) The fourth industrial revolution. Penguin, London
Searle JR (2014) What your computer can't know: review of Floridi L (2014) The fourth revolution: how the infosphere is reshaping human reality. Oxford University Press, Oxford. 9 October 2014, New York Review of Books. https://www.nybooks.com/articles/2014/10/09/what-your-computer-cant-know/. Accessed 19 July 2023
Semeraro F, Griffiths A, Cangelosi A (2023) Human–robot collaboration and machine learning: a systematic review of recent research. Robot Comput Integr Manuf 79:102432
Serota KB, Levine TR, Docan-Morgan T (2022) Unpacking variation in lie prevalence: prolific liars, bad lie days, or both? Commun Monogr 89(3):307–331
Sexton NJ, Love BC (2022) Reassessing hierarchical correspondences between brain and deep networks through direct interface. Sci Adv 8(28):eabm2219
Siemens G (2006) Connectivism: a learning theory for the digital age. Int J Instruct Technol Distance Learn 2:1–9

Silva JM et al (2022) Minimum costs to conserve 80% of the Brazilian Amazon. Perspect Ecol Conserv 20(3):216–222
Singer W (2009) Der Beobachter im Gehirn: Essays zur Hirnforschung. Suhrkamp, Frankfurt
Singer P (2019) The life you can save: how to do your part to end world poverty, 10th anniversary edn. Life You Can Save, Sydney
Šlaus I (2020) Transforming our world: necessary, urgent and still possible. Cambridge Scholars Publishing, Newcastle
Snow CP (2001 [1959]) The two cultures. Cambridge University Press, London
Soros G (2015) The alchemy of finance. Wiley, Hoboken, NJ
Spitzer M (2012) Digitale Demenz: Wie wir uns und unsere Kinder um den Verstand bringen. Droemer, Munich
Spitzer M (2019) Die Smartphone Epidemie: Gefahren für Gesundheit, Bildung und Gesellschaft. Klett-Cotta, Stuttgart
Stokes JM et al (2022) A deep learning approach to antibiotic discovery. Cell 180:688–702
Surowiecki J (2004) The wisdom of crowds: why the many are smarter than the few and how collective wisdom shapes business, economies, societies and nations. Anchor Books, New York
Tang J et al (2023) Semantic reconstruction of continuous language from non-invasive brain recordings. Nat Neurosci 26:858–866
Tegmark M (2019) Leben 3.0: Mensch sein im Zeitalter Künstlicher Intelligenz. Ullstein, Berlin
The Economist (2023) How to worry wisely about artificial intelligence. The Economist, 20 April 2023. https://www.economist.com/leaders/2023/04/20/how-to-worry-wisely-about-artificial-intelligence. Accessed 19 July 2023
Tomasello M (2019) Becoming human: a theory of ontogeny. Belknap, Cambridge, MA
Tunyasuvunakool K et al (2021) Highly accurate protein structure prediction for the human proteome. Nature 596:590–596
Turing AM (1950) Computing machinery and intelligence. Mind 59(236):433–460
Ulam S (1958) Tribute to John von Neumann. Bull Am Math Soc 64(3) Part 2:5
Utke AR (1998) Introduction: the (re)unification of knowledge: why? how? where? when? Counterpoints 39:1–33
Varela F, Thompson E (1992) Der Mittlere Weg der Erkenntnis. Scherz, Bern
Vervaeke J (2020) The meaning crisis. https://www.meaningcrisis.co. Accessed 19 July 2023
Villalobos P et al (2022) Will we run out of data? An analysis of the limits of scaling datasets in machine learning. https://arxiv.org/abs/2211.04325. Accessed 19 July 2023
von Neumann J (2000) The computer and the brain (Silliman Lectures). Yale University Press, New Haven
von Uexküll J (1957) Nie geschaute Welten. List, Munich
Vrontis D et al (2022) Artificial intelligence, robotics, advanced technologies and human resource management: a systematic review. Int J Hum Resour Manag 33(6):1237–1266
Walach H (2010) Complementary? Alternative? Integrative? Forsch Komplementmed 17(4): 215–216
Wang W, Siau K (2019) Artificial intelligence, machine learning, automation, robotics, future of work and future of humanity: a review and research agenda. J Database Manag 30(1):61–79
Wason PC (1960) On the failure to eliminate hypotheses in a conceptual task. Q J Exp Psychol 12(3):129–140
Watzlawick P (1984) The invented reality: how do we know what we believe we know? W. W. Norton, New York
Weizenbaum J (1966) ELIZA—a computer program for the study of natural language communication between man and machine. Commun ACM 9(1):36–45
Weizsäcker KF (2006) Die Geschichte der Natur: Zwölf Vorlesungen (gehalten in Göttingen 1946). Hirzel, Stuttgart
Wiki Didactic (2015) What is the meaning of complex thinking. https://edukalife.blogspot.com/2015/07/what-is-meaning-of-complex-thinking.html. Accessed 19 July 2023
Wikipedia (2023a) Culture. https://en.wikipedia.org/wiki/Culture. Accessed 19 July 2023

Wikipedia (2023b) Pugh Emerson. https://en.wikipedia.org/wiki/Emerson_Pugh. Accessed 19 July 2023
Wilber K (1995) Sex, ecology, spirituality: the spirit of evolution. Shambhala, Boston
Wilber K (1997) An integral theory of consciousness. J Conscious Stud 4(1):71–92
Wilber K (1998) The marriage of sense and soul: integrating science and religion. Random House, New York
Wilber K (2007) The integral vision: a very short introduction to the revolutionary integral approach to life, God, the universe, and everything. Shambhala, Boston
Wilber K (2022) Integral Institute. https://integralinstitute.org. Accessed 19 July 2023
Wittgenstein L (2010 [1953]) Philosophical investigations. Blackwell, Oxford
Wöhe G (2015) Einführung in die allgemeine Betriebswirtschaftslehre, 2nd edn. Pearson, Munich
Woochan K et al (2023) Therapeutic strategies of three-dimensional stem cell spheroids and organoids for tissue repair and regeneration. Bioact Mater 19:50–74
Woolley AW et al (2010) Evidence for a collective intelligence factor in the performance of human groups. Science 330(6004):686–688
Yamins D et al (2014) Performance-optimized hierarchical models predict neural responses in higher visual cortex. PNAS 111(23):8619–8624
Yong E (2022) An immense world: how animal senses reveal the hidden realms around us. Bodley Head, London

The manufacturer's authorised representative in the EU is Springer Nature Customer Service Centre GmbH, Europaplatz 3, 69115 Heidelberg, Germany. If you have any concerns regarding our products, please contact ProductSafety@springernature.com

Printed and bound by CPI Group (UK) Ltd, Croydon, CR0 4YY
01/10/2026
02225916-0005